Operator Theory
Advances and Applications
Vol. 107

Editor:
I. Gohberg

Spectral Theory of Canonical Differential Systems. Method of Operator Identities

Lev A. Sakhnovich

Springer Basel AG

Author:

Lev A. Sakhnovich
Department of Mathematics
Academy of Communication
Kuznechnaja 1
Odessa 270021
Ukraine

1991 Mathematics Subject Classification 47A68, 47A10

A CIP catalogue record for this book is available from the
Library of Congress, Washington D.C., USA

Deutsche Bibliothek Cataloging-in-Publication Data
Sachnovič, Lev A.:
Spectral theory of canonical differential systems : methods of operator
identities / Lev A. Sakhnovich. – Basel ; Boston ; Berlin : Birkhäuser, 1999
(Operator theory ; Vol. 107)
ISBN 978-3-0348-9739-6 ISBN 978-3-0348-8713-7 (eBook)
DOI 10.1007/978-3-0348-8713-7

Originally published by Birkhäuser Verlag in 1999
Softcover reprint of the hardcover 1st edition 1999
Printed on acid-free paper produced from chlorine-free pulp. TCF ∞
Cover design: Heinz Hiltbrunner, Basel

9 8 7 6 5 4 3 2 1

Contents

Introduction

The spectral theory of ordinary differential operators L and of the equations

$$Ly = \lambda y \tag{0.1}$$

connected with such operators plays an important role in a number of problems both in physics and in mathematics. Let us give some examples of differential operators and equations, the spectral theory of which is well developed.

Example 1. The Sturm-Liouville operator has the form (see [6])

$$Ly = -\frac{d^2y}{dx^2} + u(x)y = \lambda y. \tag{0.2}$$

In quantum mechanics the Sturm-Liouville operator L is known as the one-dimensional Schrödinger operator. The behaviour of a quantum particle is described in terms of spectral characteristics of the operator L.

Example 2. The vibrations of a nonhomogeneous string are described by the equation (see [59])

$$-\frac{d^2y}{dx^2} = zp^2(x)y, \qquad p(x) \geq 0. \tag{0.3}$$

The first results connected with equation (0.3) were obtained by D. Bernoulli and L. Euler. The investigation of this equation and of its various generalizations continues to be a very active field (see, e.g., [18], [19]). The spectral theory of the equation (0.3) has also found important applications in probability theory [20].

Example 3. Dirac-type systems of the form

$$\left.\begin{aligned} \frac{d\Phi}{dx} &= \frac{z}{2}\Psi + a(x)\Phi - b(x)\Psi, \\ \frac{d\Psi}{dx} &= -\frac{z}{2}\Phi - b(x)\Phi - a(x)\Psi, \end{aligned}\right\} \tag{0.4}$$

where $a(x) = \overline{a(x)}$, $b(x) = \overline{b(x)}$, are also well studied. Among the works devoted to the spectral theory of the system (0.4) the well-known article of M. G. Kreĭn [48] deserves special mention.

This book deals with the spectral theory of canonical systems of differential equations of the form

$$\frac{dY}{dx} = izJ\mathcal{H}(x)Y, \tag{0.5}$$

where J and $\mathcal{H}(x)$ are $2m \times 2m$ matrices and $Y(x)$ is a $2m \times 1$ column vector,

$$J = \begin{bmatrix} 0 & I_m \\ I_m & 0 \end{bmatrix}, \qquad \mathcal{H}(x) \geq 0. \tag{0.6}$$

(Here and throughout the book I_m denotes the identity matrix of order m.) The Sturm-Liouville equation (0.2), the string equation (0.3), and the Dirac-type system (0.4) can be transformed into a system (0.5), (0.6). Notice also that the linear Hamiltonian systems of mechanics have the form (0.5), (0.6).

Important results concerning the spectrum of canonical systems on finite intervals were obtained by I. Gohberg and M. G. Kreĭn in their book [34]. The case of canonical systems with $m = 1$ was studied e.g., in the works [13], [40].

In this book we give an exposition of the spectral theory (direct and inverse problems) of general canonical systems ($m \geq 1$).

Let us describe in more detail the contents of the book.

1. The first three chapters are devoted to factorization methods for matrix- and operator-valued functions $W(z)$. Such factorization theorems play an essential role in developing the spectral theory of canonical systems.

Factorization problems for operator-valued functions $W(z)$ can be conventionally divided into three classes.

1) Divide the complex plane Δ into two regions: Δ_+ and Δ_-. Find factors $W_+(z)$ and $W_-(z)$ such that

$$W(z) = W_+(z)W_-(z),$$

where $W_+^{\pm 1}(z)$, $W_-^{\pm 1}(z)$ have no singularities in Δ_+ and Δ_-, respectively.

2) Represent $W(z)$ as

$$W(z) = W_n(z) \cdot W_{n-1}(z) \cdot \ldots \cdot W_1(z), \qquad n \leq \infty,$$

where $W_k(z)$ are so-called *elementary factors.*

3) Represent $W(z)$ as a continual product of elementary factors.

There is a rich literature devoted to Problems 1–3 (see, e.g, [7], [28], [72]). This can be explained by the fact that factorization methods play an important role in a number of problems of analysis as well as in many applications. Problem 1) is essential for the solution of singular integral equations and of equations with kernels depending on the difference of the arguments(the Wiener-Hopf method). The solution of Problem 2) is used in the theory of electrical circuits [21], in the theory of non-selfadjoint operators [82], and in interpolation problems [37], [94]. Finally, Problem 3) is closely related to triangular models [58] and inverse spectral problems [89], [94].

Let $\mathfrak{G}$ and $\mathfrak{H}$ be Hilbert spaces. The symbol $\{\mathfrak{H}_1, \mathfrak{H}_2\}$ denotes the set of all bounded linear operators acting from $\mathfrak{H}_1$ to $\mathfrak{H}_2$.

Our investigation of factorization problems by what we call the *analytic method* is based on the *operator identity*

$$AS - SB = \Pi_1\Pi_2^*. \tag{0.7}$$

Here $A, S, B \in \{\mathfrak{H}, \mathfrak{H}\}$, $\Pi_1, \Pi_2 \in \{\mathfrak{G}, \mathfrak{H}\}$. We also introduce the operators Γ_1 and Γ_2 by mean of the relations

$$S\Gamma_1 = \Pi_1, \qquad \Gamma_2^* S = \Pi_2^*. \tag{0.8}$$

With the operator identity (0.7) one associates the *dynamical system*

$$\frac{dx}{dt} = Ax + \Pi_1 v, \qquad y = \Gamma_2^* x + v.$$

Here $y(t), v(t)$ are functions with values in $\mathfrak{G}$ and $x(t)$ is a function with values in $\mathfrak{H}$. The *transfer function* of the system (0.8) has the form

$$W_A(z) = I - \Gamma_2^*(A - zI)^{-1}\Pi_1. \tag{0.9}$$

(Here and throughout the book I denotes the identity operator in the space under consideration.) Together with $W_A(z)$ we will consider its inverse

$$W_B(z) = W_A^{-1}(z) = I + \Pi_2^*(B - zI)^{-1}\Gamma_1. \tag{0.10}$$

A representation of a given operator-valued function $W(z)$ in the form (0.9) is called a *realization*. In our approach to factorization problems an important role is played by realizations of $W(z)$ and $W^{-1}(z)$ in the form (0.9) and (0.10), respectively [82], [84]. The factorization theorem proved here has as a particular case the well-known theorem on factorization of characteristic matrix functionsAn approach to factorization problems close to ours, but based on geometric ideas, was developed in a number of publications by H. Bart, I. Gohberg and M. A. Kaashoek [7], [8], [9].

An important role in our study will be played by the notion of an *S-node*. The operators A, B, S, Π_k, Γ_k $(k = 1, 2)$ are said to form an S-node if (0.3) and (0.4) hold. This notion generalizes the well-known notion of a node [14], [58] and reduces to it if one puts

$$S = E, \qquad B = A^*, \qquad \Pi_2^* = iJ\Pi_1^*, \qquad J = J^*, \qquad J^2 = I. \tag{0.11}$$

Let us mention the following important property of S-nodes.

The equation

$$SY(z) = (A - zI)^{-1}\Pi_2 \tag{0.12}$$

has the solution

$$Y(z) = (B - zI)^{-1}\Gamma_1 W_A(z). \tag{0.13}$$

Relations (0.12), (0.13) allow us to investigate the equation $Sf = g$ with a special right-hand side g. In this way we can introduce an operator-valued function $\rho(z, \zeta)$ in terms of which one can express the *reflection* and *transmission coefficients* in *radiative transfer* problems and the *scattering diagram* in *diffraction* problems (see [5]). Our theorem on the analytic structure of the function $\rho(z,\zeta)$ generalizes considerably the well-known formula of Ambartsumyan [5].

In this book we investigate in detail the *symmetric* S-nodes, defined by (0.7) and the equalities

$$S = S^*, \quad B = A^*, \qquad \Pi_2^* = iJ\Pi_1^*, \qquad \Gamma_2^* = iJ\Gamma_1^*, \tag{0.14}$$

where

$$J = J^*, \qquad J^2 = I.$$

The operator-valued transfer function $W_A(z)$ of a symmetric S-node obeys the following *reflection principle*:

$$W_A(z)JW_A^*(\bar{z}) = J. \tag{0.15}$$

For the subsequent analysis of symmetric S-nodes and for the investigation of the connection between S and $W_A(z)$ we introduce the following notions.

A selfadjoint operator S is said to belong to the *class* $\mathcal{H}_\kappa$ if it has κ negative eigenvalues, counting multiplicities. An operator-valued function $W(z)$ is said to belong to the *class* $\mathcal{R}_\kappa$ if $W(z)$ is analytic in a neighborhood D of the point $z=\infty$ and if for any choice of vectors h_k and points z_k ($z_k \neq z_l$, $z_k \in D$), the quadratic form

$$R(\zeta_1,\dots,\zeta_n) = i\sum_{k,l=1}^{n}\left(\frac{W_A(z_k)JW_A^*(z_l)-J}{z_k-\bar{z}_l}h_l,\, h_k\right)\zeta_l\bar{\zeta}_k$$

has at most κ negative squares, and for some choice of h_k, z_k it has exactly κ negative squares (classes close to $\mathcal{R}_\kappa$ were introduced by M. G. Kreĭn and H. Langer in [51]). Under quite general assumptions we will show that the assertions $S \in \mathcal{H}_\kappa$ and $W_A(Z) \in \mathcal{R}_\kappa$ are equivalent.

We also note that in a number of important problems factorization is meaningful only in the case when the factors have certain properties inherited from the operator-valued function that one seeks to factor (see a series of papers by V. P. Potapov and his collaborators, e.g., [21]). In this direction our factorization theorem yields a factorization of $W_A(z)$ in the class $\mathcal{R}_\kappa$:

$$W_A(z) = W_2(z)W_1(z),$$

where

$$W_1(z) \in \mathcal{R}_{\kappa_1}, \qquad W_2(z) \in \mathcal{R}_{\kappa_2}, \qquad \kappa_1+\kappa_2=\kappa.$$

In particular, for $\kappa=0$ we recover the well-known Potapov factorization [21].

Under certain conditions the factorization theorem permits us to find a *continual factorization* of a given operator-valued function $W(z)$, i.e., to represent $W(z)$ as a *multiplicative integral*

$$W(z) = \int_a^{b}\!\!\!\!\!\!\curvearrowleft \exp\left[-\frac{dB(t)}{\alpha(t)-z}\right]. \tag{0.16}$$

The multiplicative integral is defined by the rule

$$\int_a^{b}\!\!\!\!\!\!\curvearrowleft \exp\left[-\frac{dB(t)}{\alpha(t)-z}\right]$$

$$= \lim_{\Delta t_j \to 0} \exp\left[\frac{\Delta_n B(t)}{\alpha(t_n)-z}\right] \cdot \exp\left[\frac{\Delta_{n-1} B(t)}{\alpha(t_{n-1})-z}\right] \cdot \ldots \cdot \exp\left[\frac{\Delta_1 B(t)}{\alpha(t_1)-z}\right],$$

where $a = t_0 < t_1 < \ldots < t_n = b$, $\Delta_k B(t) = B(t_k) - B(t_{k-1})$, and the limit is understood in the sense of matrices.

We will give a method for constructing $B(t)$ for a given operator-valued function $W(z)$.

2. The second part of the book (Chapters 4–11) is devoted to the spectral analysis of canonical systems of differential equations of the form

$$\frac{dY(x,z)}{dx} = izJ\mathcal{H}(x)Y(x,z), \qquad 0 \le x \le l \quad (l \le \infty), \tag{0.17}$$

where $\mathcal{H}(x)$ and J are $n \times n$-matrices, $Y(x,z)$ is an $n \times 1$ matrix,

$$\mathcal{H}(x) \ge 0, \qquad J = \begin{bmatrix} 0 & I_m \\ I_m & 0 \end{bmatrix}, \qquad n = 2m. \tag{0.18}$$

A number of classical spectral problems (Sturm-Liouville problems, Dirac-type systems, Kreĭn systems) can be reduced to systems of the form (0.17).

In Chapter 4 we introduce the fundamental notion of a *spectral matrix function* $\tau(u)$ of a system (0.17) with the *boundary condition*

$$D_2 Y_1(0) + D_1 Y_2(0) = 0, \tag{0.19}$$

where D_1, D_2 are $m \times m$ matrices, $Y(x,z) = \operatorname{col}[Y_1(x,z), Y_2(x,z)]$.

Let us assume that

$$D_1 D_2^* + D_2 D_1^* = 0, \qquad D_1 D_1^* + D_2 D_2^* = I_m \tag{0.20}$$

and introduce the space $L^2(\mathcal{H})$ of vector-valued functions $Y(x)$, equipped with the inner product

$$(Y(x), Y(x)) = \int_0^l Y^*(x)\mathcal{H}(x)Y(x)\,dx. \tag{0.21}$$

Denote by $\mathcal{W}(x,z)$ the matrix solution of the system

$$\frac{d\mathcal{W}(x,z)}{dx} = izJ\mathcal{H}(x)\mathcal{W}(x,z), \qquad \mathcal{W}(0,z) = I_{2m}. \tag{0.22}$$

Definition 0.1. A monotonically increasing $m \times m$ matrix-valued function $\tau(u)$ is called a *spectral matrix function* of the system (0.17), (0.19) if the operator V acting as

$$(Vg)(u) = \int_0^l [D_1 \ \ D_2]\mathcal{W}^*(x,u)\mathcal{H}(x)g(x)\,dx, \tag{0.23}$$

maps $L^2(\mathcal{H})$ isometrically into $L^2(\tau)$. (The inner product in $L^2(\tau)$ is defined by

$$(f_1, f_2)_{L^2(\tau)} = \int_{-\infty}^{\infty} f_2^*(u)[d\tau(u)]f_1(u).\ )$$

Definition 0.2. A spectral matrix function $\tau(u)$ of the system (0.17), (0.19) is said to be *orthogonal* if the operator V maps $L^2(\mathcal{H})$ unitarily *onto* $L^2(\tau)$.

Spectral matrix functions $\tau(u)$ can be also introduced by resorting to the general spectral theory of operators (see [2]). In this book we do not adopt such an aproach, though the ideas of the general spectral theory have certainly served us as a guide. In the case of a finite interval we prove (see Chapter 4, Corollary 2.2) that the equality $L = V^{-1}QV$ holds. Here L is the operator generated by the canonical system and the boundary conditions imposed and the operator Q acts according to the rule $(Qf)(u) = uf(u)$ for $f \in L^2(\tau)$. Thus, the notion of spectral matrix-function $\tau(u)$ introduced above coincides with the notion of the spectral matrix-function in the general theory of operators (see [2]). In Chapters 4–11 we consider both direct and inverse spectral problems for systems (0.17), (0.19). The *direct spectral problem* can be formulated as follows.

Given a canonical system (0.17), (0.19), *describe the set of all spectral matrix functions* $\tau(u)$ *of this system.*

The *inverse spectral problem* is:

Given a monotonically increasing $m \times m$ *matrix-valued function* $\tau(u)$, *find a canonical system* (0.17), (0.19) *for which* $\tau(u)$ *is a spectral matrix function.*

We will consider here spectral problems for canonical systems of the following types:

I. Problems on a finite interval $0 \le x \le l$.

II. Problems on the half-line $0 \le x < \infty$.

III. Problems on the line $-\infty < x < \infty$.

In Chapters 4–11 we describe the connections among these problems and specific features of each of them. The main attention is devoted to the problems on the half-line and on the line.

We will see that the investigation of problems with the boundary condition (0.19) can be reduced to the case of the boundary condition

$$Y_1(0) = 0, \tag{0.24}$$

i.e., to the case $D_1 = 0$, $D_2 = I_m$.

An important role in the spectral theory of systems (0.17), (0.24) is played by the $m \times m$ matrix-valued functions $v(z)$ defined by the following relation:

$$v(z) = i\left[a(z)R(z) + b(z)Q(z)\right] \cdot \left[c(z)R(z) + d(z)Q(z)\right]^{-1}, \qquad \operatorname{Im} z > 0. \quad (0.25)$$

Here $R(z), Q(z)$ are $m \times m$ matrix-valued functions meromorphic in the upper half-plane such that

$$\det[R^*(z)R(z) + Q^*(z)Q(z)] \not\equiv 0, \qquad \operatorname{Im} z > 0, \quad (0.26)$$

$$R^*(z)Q(z) + Q^*(z)R(z) \geq 0, \qquad \operatorname{Im}\ z > 0. \quad (0.27)$$

The matrix of coefficients of the linear-fractional transformation (0.25) is given in terms of the solution $\mathcal{W}(x, z)$ (see (0.22)) by

$$\mathcal{U}(z) = \begin{bmatrix} a(z) & b(z) \\ c(z) & d(z) \end{bmatrix} = \mathcal{W}^*(l, z), \quad (0.28)$$

where $a(z), b(z), c(z), d(z)$ are $m \times m$ matrix-valued functions.

The matrix-valued functions $v(z)$ defined by formulas (0.25)–(0.28) are analogues of the classical Weyl-Titchmarsh functions encountered in the theory of Sturm-Liouville equations. They have the remarkable property of belonging to the *Nevanlinna class*, i.e.,

$$\frac{v(z) - v^*(z)}{i} \geq 0,$$

and consequently admit a representation

$$v(z) = \alpha + \int_{-\infty}^{\infty} \left(\frac{1}{u - z} - \frac{u}{1 + u^2} \right) d\tau(u), \quad (0.29)$$

where $\alpha = \alpha^*$ and $\tau(u)$ is a monotonically increasing $m \times m$ matrix-valued function.

If the system (0.17) satisfies some additional conditions the following result holds (Chapter 4):

The set of all spectral matrix functions of the system (0.17) *coincides with the set of all matrix-valued functions* $\tau(u)$ *that arise in representations* (0.29) *of matrix-valued functions* $v(z)$ *of type* (0.25), (0.28).

Hence, α and $\tau(u)$ from representation (0.29) are spectral characteristics of the canonical system (0.17), (0.24).

We see that formulas (0.25)–(0.29) provide the solution of the direct spectral problem, i.e., we obtain a description of the set of all spectral characteristics $\tau(u)$ and α of the system (0.17), (0.24).

Next, let us describe some of the results concerning the inverse spectral problem. Here we are given the spectral characteristics $\tau(u)$ and α, and we are required to reconstruct systems (0.17) with these spectral characteristics.

An essential role in solving the inverse spectral problem is played by the operator identity

$$AS - SA^* = i(\Phi_1\Phi_2^* + \Phi_2\Phi_1^*), \tag{0.30}$$

where the operators Φ_1, Φ_2 act from the Hilbert space $\mathfrak{G}_1$ ($\dim \mathfrak{G}_1 < \infty$) into $\mathfrak{H}$. Identity (0.30) is a special case of identity (0.7), obtained when

$$\Pi_1 = [\Phi_1 \;\; \Phi_2], \qquad \Pi_2 = [-i\Phi_2 \;\; -i\Phi_1], \qquad \mathfrak{G} = \mathfrak{G}_1 \oplus \mathfrak{G}_2.$$

We assume that there exists a continuous monotonically increasing set of orthogonal projectors P_ζ ($0 \leq \zeta \leq \infty$) such that

$$A^*P_\zeta = P_\zeta A^* P_\zeta. \tag{0.31}$$

Our recipe for solving the inverse spectral problem is as follows. Given the spectral data $\tau(u)$ and α, pick some fixed operators A and Φ_2. Now find the operators S and Φ_1 by means of the formulas

$$S = \int_{-\infty}^{\infty} (I - Au)^{-1}\Phi_2[d\tau(u)]\Phi_2^*(I - A^*u)^{-1}, \tag{0.32}$$

$$\Phi_1 = -i \int_{-\infty}^{\infty} \left[A(I - Au)^{-1} + \frac{u}{1 + u_2}\right] \Phi_2 \, d\tau(u) + i\Phi_2\alpha. \tag{0.33}$$

Finally, set

$$\mathcal{H}(x) = B'(x), \tag{0.34}$$

where

$$B(\zeta) = \Pi^* S_\zeta^{-1} P_\zeta \Pi, \qquad \Pi = [\Phi_1 \;\; \Phi_2], \qquad S_\zeta = P_\zeta S P_\zeta.$$

This completes the solution of the inverse spectral problem.

We will denote by $\mathbf{N}(A, \Phi_2)$ the class of systems (0.17) defined by formulas (0.32)–(0.34).

The method described above for solving the inverse problem is based on results of a general theory of interpolation ([94], Chapter 1).

In the case of the Sturm-Liouville equation (0.2) the method of solving the inverse problem defined by the recipe (0.30)–(0.34) is in fact identical to the classical Gelfand-Levitan method [26]. Also, in the case of the Dirac-type system (0.4) our method coincides with the classical method of M. G. Kreĭn [48].

In Chapter 5 we investigate the direct and inverse spectral problems for canonical systems on the line $(-\infty, \infty)$. As in the case of the Sturm-Liouville equation, the problem on the line can be reduced to the problem on the half-line $(0, \infty)$ by doubling the dimension of the system (see [56], Chapter 7). The spectral and scattering problems on the line for the Sturm-Liouville equation are described, e.g., in the books [56], [59].

Chapter 6 is devoted to an important special case of problems on the line, namely, the *periodic problem*, when the condition

$$\mathcal{H}(x+l) = \mathcal{H}(x), \qquad -\infty < x < \infty, \tag{0.35}$$

is satisfied. Therein we conduct a more detailed study of the Weyl-Titchmarsh matrix function $v(z)$. The character of the singularities of $v(z)$ (branch points) is determined by the Jordan structure of the monodromy matrix

$$W(z) = \mathcal{W}(l, z). \tag{0.36}$$

In Chapter 7 (Section 1) we consider the case when the spectral matrix functions $\widetilde{\tau}(u)$ and $\tau(u)$ of two canonical systems are connected by the relation

$$\widetilde{\tau}(u) = \tau(u) + \gamma(u), \tag{0.37}$$

where $\gamma(u)$ is a piecewise constant matrix-valued function. The operators A and Φ_2 are fixed. It is proved that the Hamiltonians $\widetilde{\mathcal{H}}(x)$ and $\mathcal{H}(x)$ constructed according to the recipe (0.30)–(0.34) are such that the matrices $J\widetilde{\mathcal{H}}(x)$ and $J\mathcal{H}(x)$ are linearly similar for all x.

Let us emphasize that in the case (0.37) we provide formulas for the transition from the Hamiltonian $\mathcal{H}(x)$ to the Hamiltonian $\widetilde{\mathcal{H}}(x)$, i.e., we solve the inverse problem for the case (0.37) in explicit form. Further (Section 2) it is proved that the theorem on similarity of $J\widetilde{\mathcal{H}}(x)$ and $J\mathcal{H}(x)$ remains valid when the assumptions on $\gamma(u)$ are relaxed.

In Chapter 8 we single out the classes of canonical systems for which the inverse problem has a unique solution. In his report at the International Congress of Mathematicians (Moscow, 1966, see [49]) M. G. Kreĭn stated: "It looks as if the problem of finding additional conditions under which the Hamiltonian $\mathcal{H}(x)$ can be uniquely determined is extremely difficult and it is unlikely that it will be solved in the near future".

Kreĭn's words proved prophetic. Indeed, little progress has been achieved in this problem in the thirty years that passed since his report. Uniqueness theorems have been proved only for classical systems (matrix Sturm-Liouville equations, Dirac-type systems) and for systems close to the classical ones.

In Chapter 8 we establish uniqueness theorems in two different situations. In the first, under some assumptions, we prove uniqueness theorems when the monodromy matrix $W(z) = \mathcal{W}(l, z)$ is known. Notice that in Chapter 3 we give a method of constructing a canonical system when one knowns the monodromy matrix $W(z)$. The second situation concerns the central problem of this book, namely, when the spectral data $\tau(u)$ and α are known. Conditions under which the corresponding inverse problem has a unique solution are found.

In Chapter 9 we continue to investigate the matrix-valued functions $v(l, z)$ defined by formulas (0.25)–(0.28). The matrices $v(l, z)$ form a system of *nested*

Weyl discs. Conditions under which the limit Weyl disc ($l \to \infty$) reduces to a point (the *Weyl limit point*) are given. We prove theorems that generalize the well-known results obtained by B. M. Levitan [57] for Dirac-type systems. Notice that the Weyl-Titchmarsh matrix-valued functions $v(l, z)$ are introduced earlier, in Chapter 4, in connection with the spectral characteristics of the canonical system of differential equations (0.17). In Chapter 9 they reappear in connection with the problem of the number of linearly independent solutions of a system (0.17) that belong to the space $L^2(\mathcal{H})$ ($l = \infty$). The point is that the fact that the Weyl-Titchmarsh matrix-valued function $v(z)$ belongs to the Weyl limit disc can be expressed by the inequality

$$\int_0^\infty [I_m \quad iv^*(z)]\mathcal{W}^*(x,z)\mathcal{H}(x)\mathcal{W}(x,z)\begin{bmatrix} I_m \\ -iv(z)\end{bmatrix} dx < \infty. \tag{0.38}$$

In Chapter 9 we also consider the canonical system (0.17) on the line ($-\infty < x < \infty$) and introduce the corresponding Weyl-Titchmarsh matrix-valued function. Together with the periodic system (0.17),

$$\mathcal{H}(x+l) = \mathcal{H}(x), \qquad -\infty < x < \infty,$$

we consider the system with shifted argument

$$\frac{dY}{dx} = izJ\mathcal{H}(x+\tau)Y, \qquad -\infty < x < \infty \tag{0.39}$$

and we exhibit the relationship between the Weyl-Titchmarsh function $v(\tau, z)$ of the system (0.39) and $v(z) = v(0, z)$. In particular, we show that the branch points of $v(z)$ are invariant under the shift of the argument. A similar result for the Sturm-Liouville periodic problem was obtained by B. M. Levitan and used by him in the analysis of the periodic solutions of the Korteweg-de Vries equation [56].

In Chapter 10 the general spectral theory of canonical systems developed in chapters 4–9 is used to investigate concrete classes of systems. Specifically, we analyze canonical systems (0.17) for which the matrix $J\mathcal{H}(x)$ is linearly similar to the matrix

$$M_1 = \begin{bmatrix} D & 0 \\ 0 & 0 \end{bmatrix}, \qquad D = \operatorname{diag}\{d_1, d_2, \dots, d_m\}, \qquad d_k > 0,$$

i.e.,

$$J\mathcal{H}(x) = T(x)M_1T^{-1}(x). \tag{0.40}$$

When $m > 1$ and $d_j \neq d_k$ ($j \neq k$) the system (0.17) generates nonclassical spectral problems, which play an essential role both in theoretical and applied domains. For $m = 1$ systems of the form (0.17), (0.40) were introduced and investigated in a well-known paper by M. G. Kreĭn [48]. Unfortunately, Kreĭn did not publish complete proofs of the results formulated in that paper. Here we generalized a number of

Kreĭn's results given in [48] for the case $m \geq 1$ and give the corresponding proofs. In Chapter 10 we focus on the classical Dirac-type system

$$\begin{aligned} \frac{d\Phi}{dx} &= \frac{z}{2}\Psi + a(x)\Phi - b(x)\Psi, \\ \frac{d\Psi}{dx} &= -\frac{z}{2}\Phi - b(x)\Phi - a(x)\Psi, \end{aligned}$$

where $a(x) = a^*(x)$, $b(x) = b^*(x)$. The Dirac-type system is obtained from the system (0.17), (0.40) if one sets

$$d_1 = d_2 = \ldots = d_m = 1.$$

In Chapter 11 the general spectral theory developed in Chapter 4 is used to study the matrix string equation

$$P^{-1}(x)\frac{d}{dx}\left\{A(x)\frac{d}{dx}P^{-1}(x)Y(x,z)\right\} = zY(x,z), \tag{0.41}$$

as well as the matrix Sturm-Liouville equation

$$-\frac{d^2Y}{dx^2} + U(x)Y = zY, \tag{0.42}$$

where $P(x)$, $A(x)$ and $U(x)$ are $m \times m$ matrices.

A description of the set of the spectral matrix functions of systems (0.41) and (0.42) is given, i.e., the direct spectral problem is solved. The inverse spectral problem is also solved under additional conditions, i.e., a method of reconstructing the systems (0.41) and (0.42) from their spectral data is given.

3. In Chapter 12, the last chapter of the book, the results of the spectral theory are used to investigate some integrable nonlinear partial differential equations (the nonlinear Schrödinger equation, the modified Korteweg-de Vries equation, the sinh-Gordon equation).

In 1967 Gardner, Green, Kruskal, and Miura [23] discovered a new approach to the analysis of the nonlinear Korteweg-de Vries equation

$$U_t - 6UU_x + U_{xxx} = 0. \tag{0.43}$$

Specifically, they used results on the inverse scattering problem for the equation

$$-\frac{d^2y}{dx^2} + u(x,t)y = \lambda y, \qquad -\infty < x < \infty.$$

P. Lax [24] and V. E. Zakharov and A. B. Shabat [98] developed further what has become known as the *method of the inverse scattering problem*. It soon became clear that this method works not only for the Korteweg-de Vries equation, but also for a large number of other nonlinear equations of mathematical physics. The method of the inverse scattering problem has been widely applied by many authors and continues to be a very active field of research (see, e.g., [1], [22]).

The method of the inverse spectral problem was for the first time applied to the analysis of nonlinear equations by Yu. M. Berezanskiĭ [11] (in the discrete case), and by the author of this book [86], [88] (in the discrete and continuous cases).

The transition from the inverse scattering problem to the inverse spectral problem removes the requirement of regularity of solutions at infinity, and enables one to construct new classes of explicit solutions for a number of nonlinear equations.

Also, the method of the inverse spectral problem method allows us to investigate relations between boundary conditions and initial conditions for a number of integrable nonlinear equations.

Chapter 12 (Section 1) is devoted to a description of the main ideas of the method of the inverse spectral problem. For this purpose we consider the so-called zero-curvature equation (see [15], [22])

$$\frac{\partial G(x,t,z)}{\partial t} - \frac{\partial F(x,t,z)}{\partial x} + G(x,t,z)F(x,t,z) - F(x,t,z)G(x,t,z) = 0, \quad (0.45)$$

where $G(x,t,z), F(x,t,z)$ are $2m \times 2m$ matrix-valued functions and $G(x,t,z)$ has the form

$$G(x,t,z) = izJ\mathcal{H}(x,t). \quad (0.46)$$

With the nonlinear partial differential equation (0.45) one can associate the canonical system (parametrized by the variable t)

$$\frac{\partial \mathcal{W}(x,t,z)}{\partial x} = G(x,t,z)\mathcal{W}(x,t,z), \qquad 0 \le x < \infty, \qquad t \ge 0. \quad (0.47)$$

The main result of Section 1, Chapter 12 yields a formula which expresses the Weyl-Titchmarsh matrix function $v(t,z)$ of the system (0.47) in terms of $v(z) = v(0,z)$ and $F(0,t,z)$ (*evolution law*). In Sections 2 and 3 of Chapter 12 the general result is applied to the investigation of concrete nonlinear equations: the nonlinear Schrödinger equation, the modified Korteweg-de Vries equation, and the sinh-Gordon equation. In particular, in Section 3 we find conditions ensuring the uniqueness of the solution of the listed equations. In Section 4, Chapter 12 we construct a hierarchy of matrix nonlinear partial differential equations

$$\frac{\partial G(x,t,z)}{\partial t} - \frac{\partial F_N(x,t,z)}{\partial x} + G(x,t,z)F_N(x,t,z) - F_N(x,t,z)G(x,t,z) = 0, \quad N = 1,2,3,\dots . \quad (0.48)$$

Here $G(x,t,z)$ and $F_N(x,t,z)$ are $2m \times 2m$ matrix-valued functions of the form

$$G(x,t,z) = g_1(x,t) + zg_0, \tag{0.49}$$

$$F_N(x,t,z) = f_N(x,t) + zf_{N-1}(x,t) + \ldots + z^N f_0, \tag{0.50}$$

where

$$g_0 = f_0 = \begin{bmatrix} I_m & 0 \\ 0 & -I_m \end{bmatrix}. \tag{0.51}$$

When $N = 2$ equation (0.48) can be reduced to the matrix nonlinear Schrödinger equation, while when $N = 3$ it can be reduced to the matrix modified Korteweg-de Vries equation.

In Chapter 12 (Sections 1–3) we show how to apply the linear spectral theory to the analysis of nonlinear equations.

It is worth mentioning that the results of the analysis of nonlinear equations (0.48) in their turn allow one to obtain results in the linear theory (for instance, the asymptotics of the Weyl-Titchmarsh function $v(z)$ in the neighbourhood of the point $z = \infty$) (see 12.4).

In order to read this book it suffices if the reader is familiar with the basic notions of operator theory (see, e.g., the books [2] or [29]) and the classical results on the Sturm-Liouville problem (see, e.g., the books [16], [57] or [59]).

I am deeply grateful to I. Gohberg for his attention, help and careful edition of the book. I am also very grateful to H. Dym for a number of very valuable remarks referring to Chapter 10. Thanks are due to A. Sakhnovich for reading the book and making many useful remarks. I also wish to thank my wife E. Melnichenko for translating this book into English.

Chapter 1
Factorization of Operator-valued Transfer Functions

Consider a dynamical system of the form

$$\frac{dx}{dt} = Ax + \Pi_1 v, \qquad y = \Gamma_2^* x + v, \tag{0.1}$$

where $v(t)$ and $y(t)$ are functions taking values in a Hilbert space $\mathfrak{G}$, and $x(t)$ is a function taking values in a Hilbert space $\mathfrak{H}$. We will assume that

$$A \in \{\mathfrak{H}, \mathfrak{H}\}, \qquad \Pi_1 \in \{\mathfrak{G}, \mathfrak{H}\}, \qquad \Gamma_2^* \in \{\mathfrak{H}, \mathfrak{G}\}.$$

(The symbol $\{\mathfrak{H}_1, \mathfrak{H}_2\}$ denotes the set of all bounded linear operators acting from $\mathfrak{H}_1$ to $\mathfrak{H}_2$.) The derivative in the left-hand side of (0.1) is understood in the following sense:

$$\left\| \frac{x(t+\Delta t) - x(t)}{\Delta t} - \frac{dx}{dt} \right\| \to 0 \quad \text{for } \Delta t \to 0.$$

It is known that the operator-valued transfer function of the system (0.1) has the form

$$W_A(z) = I - \Gamma_2^*(A - zI)^{-1}\Pi_1. \tag{0.2}$$

A representation of a given operator-valued function in the form (0.2) is called a *realization.*

A typical feature of our approach is that together with the realization (0.2) we use the realization

$$W_A^{-1}(z) = W_B(z) = I + \Pi_2^*(B - zI)^{-1}\Gamma_1\,, \tag{0.3}$$

where

$$B \in \{\mathfrak{H}, \mathfrak{H}\}, \qquad \Gamma_1 \in \{\mathfrak{G}, \mathfrak{H}\}, \qquad \Pi_2^* \in \{\mathfrak{H}, \mathfrak{G}\}.$$

A realization (0.2) is said to be *minimal* if

$$\overline{\bigvee_{n=0}^{\infty} A^n \Pi_1 \mathfrak{G}} = \mathfrak{H}, \qquad \overline{\bigvee_{n=0}^{\infty} A^{*n} \Gamma_2 \mathfrak{G}} = \mathfrak{H}. \tag{0.4}$$

Section 1 of this chapter is devoted to conditions for realization of transfer functions $W(z)$. In particular, it contains a simple derivation of Kalman's theorems [43] on realization of rational matrix-valued functions. The main problem of the present chapter, the factorization problem of $W(z)$, is considered in Section 2.

Proceeding from realizations of $W(z)$ and $W^{-1}(z)$, we succeed in finding conditions for factorization and also in deriving formulas for factors. In Section 3 the general theorems are applied to the investigation of rational operator-valued functions, in particular, of operator pencils.

1.1. Realization of operator-valued functions

1.1.1. We begin by obtaining a realization of $W_A^{-1}(z)$. The following assertion holds [82].

Theorem 1.1. *Suppose that an operator-valued function $W_A(z)$ admits a realization (0.2). If z does not belong to the spectra of A and of*

$$B = A - \Pi_1\Gamma_2^*, \tag{1.1}$$

then

$$W_A^{-1}(z) = W_B(z) = I + \Gamma_2^*(B - zI)^{-1}\Pi_1. \tag{1.2}$$

Proof. Consider the product

$$W_A(z)W_B(z) = [I - \Gamma_2^*(A - zI)^{-1}\Pi_1]\,[I + \Gamma_2^*(B - zI)^{-1}\Pi_1]. \tag{1.3}$$

Since $\Pi_1\Gamma_2^* = A - B = (A - zI) - (B - zI)$, we deduce from relation (1.3) that $W_A(z)W_B(z) = I$. In a similar way one can show that $W_B(z)W_A(z) = I$. This completes the proof of the theorem. □

Proposition 2. *If an operator-valued function $W(z)$ admits a realization, then $W(z)$ admits also a minimal realization.*

1.1.2. Let us give some examples of operator-valued functions admitting a realization. These examples play an essential role in subsequent constructions.

Example 1. Let $W(z)$ be an $m \times m$ matrix-valued function ($m < \infty$) of the form

$$W(z) = I - \frac{D}{a - z}, \qquad \text{rank } D = 1. \tag{1.4}$$

By (1.4), D admits a representation

$$D = \text{col}\,[\Gamma_1, \Gamma_2, \dots, \Gamma_m] \cdot [p_1, p_2, \dots, p_m]. \tag{1.5}$$

Then $W(z)$ admits a realization (0.2), where

$$Ae = ae, \qquad e \in \mathfrak{H}, \qquad \dim \mathfrak{H} = 1, \tag{1.6}$$

$$\Gamma_2^* e = \text{col}\,[\Gamma_1, \Gamma_2, \dots, \Gamma_m]e, \qquad \Pi_1 g = [p_1, p_2, \dots, p_m]g, \tag{1.7}$$

and $g = \text{col}\,[g_1, g_2, \dots, g_m] \in \mathfrak{G}$.

Example 2. To obtain a realization of the function $W(z) = \exp(\gamma/z)$, $\dim \mathfrak{G} = 1$, consider the operator

$$Af = \gamma \int_0^x f(t)\,dt, \qquad f \in L^2(0,1) = \mathfrak{H}. \tag{1.8}$$

Then

$$(A - zI)^{-1}1 = -\frac{1}{z}\exp(\gamma x/z). \tag{1.9}$$

Now put

$$\Pi_1 1 = 1, \qquad \gamma_2^* f = \Gamma \int_0^1 f(t)\,dt. \tag{1.10}$$

By (1.9) and (1.10), one has the realization

$$\exp(\gamma/z) = 1 - \Gamma_2^*(A - zI)^{-1}\Pi_1. \tag{1.11}$$

Example 3. Let $W(z)$ be an operator pencil of the form

$$W(z) = I + \frac{1}{z}A_1 + \ldots + \frac{1}{z^m}A_m, \tag{1.12}$$

where $A_k \in \{\mathfrak{G}, \mathfrak{G}\}$, $1 \le k \le m$. We take for $\mathfrak{H}$ the orthogonal sum of m copies of $\mathfrak{G}$ and write the operators $A, B \in \{\mathfrak{H}, \mathfrak{H}\}$ in the block-matrix form

$$A = \begin{bmatrix} 0 & 0 & \ldots & 0 & 0 \\ I & 0 & \ldots & 0 & 0 \\ 0 & I & \ldots & 0 & 0 \\ \ldots & \ldots & \ldots & \ldots & \ldots \\ 0 & 0 & \ldots & I & 0 \end{bmatrix}, \quad B = \begin{bmatrix} -A_1 & -A_2 & \ldots & -A_{m-1} & -A_{m-1} \\ I & 0 & \ldots & 0 & 0 \\ 0 & I & \ldots & 0 & 0 \\ \ldots & \ldots & \ldots & \ldots & \ldots \\ 0 & 0 & \ldots & I & 0 \end{bmatrix}. \tag{1.13}$$

A direct computation yields the realizations

$$W(z) = I + P_1^* B(A - zI)^{-1}P_1, \tag{1.14}$$

$$W^{-1}(z) = I - P_1^* B(B - zI)^{-1}P_1, \tag{1.15}$$

where the operator $P_1 \in \{\mathfrak{G}, \mathfrak{H}\}$ is defined by

$$P_1 g = \operatorname{col}\,[g, 0, \ldots, 0], \quad g \in \mathfrak{G}. \tag{1.16}$$

Remark 1.1. The realizations (1.14), (1.15) are minimal if and only if the range of A_m^* is dense in $\mathfrak{G}$.

Example 4. Let $W(z)$ be an operator-valued function of the form

$$W(z) = E + \sum_{k=1}^{m} \sum_{i=1}^{m_k} \frac{D_{ki}}{(z_k - z)^i}, \tag{1.17}$$

where $D_{ki} \in \{\mathfrak{G}, \mathfrak{G}\}$, $\dim \mathfrak{G} < \infty$. It readily verified that $W(z)$ admits a representation

$$W(z) = W_2(z) W_1(z), \tag{1.18}$$

where $W_2(z)$ is a pencil of the form (1.12) and

$$W_1^{-1}(z) = (1 + a_1 z^{-1} + \ldots + a_n z^{-n}) I \tag{1.19}$$

is a scalar operator-valued function. The operator-valued functions $W_1^{-1}(z)$ and $W_2(z)$ admit realizations (Example 3). According to Theorem 1.1, $W_1(z)$ also admits a realization. We conclude that the operator-valued function $W(z)$ of the form (1.17) admits a realization. Indeed, the following result holds true.

Theorem 1.2. *Let $W_k(z)$ $(k = 1, 2)$ be operator-valued functions acting in a Hilbert space $\mathfrak{G}$ and admitting realizations*

$$W_k(z) = I - \Gamma_2^{*(k)} (A_k - zI)^{-1} \Pi_1^{(k)}, \tag{1.20}$$

where $A_k \in \{\mathfrak{H}_k, \mathfrak{H}_k\}$, $\Pi_1^{(k)}, \Gamma_2^{(k)} \in \{\mathfrak{G}, \mathfrak{H}_k\}$. Then

$$W(z) = W_2(z) W_1(z) = I - \Gamma_2^* (A - zI)^{-1} \Pi_1, \tag{1.21}$$

with $\mathfrak{H} = \mathfrak{H}_1 \oplus \mathfrak{H}_2$, $\Gamma_2 = \Gamma_2^{(1)} + \Gamma_2^{(2)}$, $\Pi_1 = \Pi_1^{(1)} + \Pi_1^{(2)}$,

$$A = \begin{bmatrix} A_1 & 0 \\ A_{21} & A_2 \end{bmatrix}, \qquad A_{21} = \Pi_1^{(2)} \Gamma_2^{(1)*}. \tag{1.22}$$

The validity of (1.21) can be verified by direct calculation.

Corollary 1.1. *Any rational operator-valued function $W(z)$ of form (1.17) admits a realization with $\dim \mathfrak{H} < \infty$.*

This fact was originally established by R. E. Kalman (see [43]) using a more complicated approach based of the theory of modules.

1.1.3. In representation (0.2) the operators A, Π_1, Γ_2 are not uniquely determined. Obviously, the following holds.

Proposition 3. *Let $W_A(z)$ be of the form (0.2). If V is a bounded operator in $\{\mathfrak{H}, \widetilde{\mathfrak{H}}\}$ with bounded inverse, then*

$$W_A(z) = W_{\widetilde{A}}(z) = I - \widetilde{\Gamma}_2^* (\widetilde{A} - zI)^{-1} \widetilde{\Pi}_1, \tag{1.23}$$

where

$$\widetilde{A} = V A V^{-1}, \qquad \widetilde{\Gamma}_2^* V = \Gamma_2^*, \qquad \widetilde{\Pi}_1 = V \Pi_1 \tag{1.24}$$

The converse is true only in the finite-dimensional case. Namely, the following holds.

Theorem 1.3 (Kalman [43]). *Suppose there are given two minimal realizations of an operator-valued function* $W(z)$:

$$W(z) = I - \Gamma_2^*(A - zI)^{-1}\Pi_1, \qquad W(z) = I - \widetilde{\Gamma}_2^*(\widetilde{A} - zI)^{-1}\widetilde{\Pi}_1, \tag{1.25}$$

where $A \in \{\mathfrak{H}, \mathfrak{H}\}$, $\widetilde{A} \in \{\widetilde{\mathfrak{H}}, \widetilde{\mathfrak{H}}\}$, $\Pi_1, \Gamma_2 \in \{\mathfrak{G}, \mathfrak{H}\}$, $\widetilde{\Pi}_1, \widetilde{\Gamma}_2 \in \{\mathfrak{G}, \widetilde{\mathfrak{H}}\}$. *If*

$$\dim \mathfrak{H} < \infty, \tag{1.26}$$

then there exists a bounded operator $V \in \{\mathfrak{H}, \widetilde{\mathfrak{H}}\}$ *with bounded inverse such that relations* (1.24) *hold.*

Proof. Putting $B = A - \Pi_1\Gamma_2^*$ and using formulas (1.2) we have

$$\begin{aligned} W^{-1}(\zeta)W(z) &= [I + \Gamma_2^*(B - \zeta I)^{-1}\Pi_1][I - \Gamma_2^*(A - zI)^{-1}\Pi_1] \\ &= I + (\zeta - z)\Gamma_2^*(B - \zeta I)^{-1}(A - zI)^{-1}\Pi_1. \end{aligned} \tag{1.27}$$

A relation similar to (1.27) is also valid for the realization of $W(z)$ in $\widetilde{\mathfrak{H}}$ with $\widetilde{B} = \widetilde{A} - \widetilde{\Pi}_1\Gamma_2^*$. Therefore,

$$\Gamma_2^*(B - \zeta I)^{-1}(A - zI)^{-1}\Pi_1 = \widetilde{\Gamma}_2^*(\widetilde{B} - \zeta I)^{-1}(\widetilde{A} - zI)^{-1}\widetilde{\Pi}_1. \tag{1.28}$$

Expanding both sides of (1.28) in series of negative powers of ζ and z, we deduce that

$$(A^k\Pi_1 g, B^{*l}\Gamma_2 f) = (\widetilde{A}^k\widetilde{\Pi}_1 g, \widetilde{B}^{*l}\widetilde{\Gamma}_2 f) \tag{1.29}$$

for all $g, f \in \mathfrak{G}$ and all $k, l = 0, 1, 2, \ldots$. Since the realizations (1.25) are minimal,

$$\overline{\bigvee_{n=0}^{\infty} A^n\Pi_1\mathfrak{G}} = \overline{\bigvee_{n=0}^{\infty} B^{*n}\Gamma_2\mathfrak{G}} = \mathfrak{H}, \qquad \overline{\bigvee_{n=0}^{\infty} \widetilde{A}^n\widetilde{\Pi}_1\mathfrak{G}} = \overline{\bigvee_{n=0}^{\infty} \widetilde{B}^{*n}\widetilde{\Gamma}_2\mathfrak{G}} = \widetilde{\mathfrak{H}}. \tag{1.30}$$

Define the operators V and U by

$$V\left(\sum_{k=0}^{s} A^k\Pi_1 g_k\right) = \sum_{k=0}^{s} \widetilde{A}^k\widetilde{\Pi}_1 g_k, \qquad g_k \in \mathfrak{G}, \qquad s < \infty \tag{1.31}$$

and

$$U\left(\sum_{k=0}^{s} B^{*k}\Gamma_2 g_k\right) = \sum_{k=0}^{s} \widetilde{B}^{*k}\widetilde{\Gamma}_2 g_k \qquad g_k \in \mathfrak{G}, \qquad s < \infty. \tag{1.32}$$

By (1.29), the equality $\sum_{k=0}^{s} A^k\Pi_1 g_k = 0$ implies that $\sum_{k=0}^{s} \widetilde{A}^k\widetilde{\Pi}_1 g_k = 0$. Hence, by relation (1.31), V associates with each element of $\mathfrak{H}$ a unique element of $\widetilde{\mathfrak{H}}$. Similarly, by (1.32), U associates with each element of $\mathfrak{H}$ a unique element of $\widetilde{\mathfrak{H}}$.

It follows from (1.26) and (1.30)–(1.32) that $\dim \widetilde{\mathfrak{H}} < \infty$ and that V and U are linear. Next, by (1.29) and (1.31), (1.32),

$$U^*V = I. \tag{1.33}$$

For $s = 0$ it follows from (1.31) and (1.32) that

$$V\Pi_1 = \widetilde{\Pi}_1, \qquad U\Gamma_2 = \widetilde{\Gamma}_2. \tag{1.34}$$

Further, the first relation in (1.34) and relation (1.31) imply that

$$VA = \widetilde{A}V. \tag{1.35}$$

The equalities (1.24) that we have to prove follow directly from (1.33)–(1.35). □

Let us emphasize again that the original proof of Theorem 1.2 due to Kalman, like that of Corollary 1.1, was based on the theory of modules and was essentially more complicated than the proof given here.

1.2. A factorization method

1.2.1. Let $W(z)$ be an operator-valued function that acts in a Hilbert space $\mathfrak{G}$ and admits realizations (0.2), (0.3). The method presented below reduces the factorization problem to a number of simpler problems. An essential role here is played by the operator S defined by the equation

$$AS - SB = \Pi_1\Pi_2^* \tag{2.1}$$

and the conditions

$$S\Gamma_1 = \Pi_1, \qquad \Gamma_2^*S = \Pi_2^*. \tag{2.2}$$

Suppose that the space $\mathfrak{H}$ in which the operators A, B, S act is represented in the form

$$\mathfrak{H} = \mathfrak{H}_1 \oplus \mathfrak{H}_2. \tag{2.3}$$

Write a vector $f \in \mathfrak{H}$ in the form $f = \operatorname{col}[f_1, f_2]$, $f_k \in \mathfrak{H}_k \quad (k = 1, 2)$. Then the operators A, B, S admit the block representations

$$A = \{A_{ij}\}_{i,j=1}^2, \qquad B = \{B_{ij}\}_{i,j=1}^2, \qquad S = \{S_{ij}\}_{i,j=1}^2.$$

Lemma 2.1. *If S_{11} is invertible on $\mathfrak{H}_1$, then S admits the factorization*

$$S = VU, \tag{2.4}$$

where V and U have block-triangular structure:

$$V = \begin{bmatrix} V_{11} & 0 \\ V_{21} & V_{22} \end{bmatrix}, \qquad U = \begin{bmatrix} U_{11} & U_{12} \\ 0 & U_{22} \end{bmatrix}. \tag{2.5}$$

Equality (2.4) holds for the following choice of elements:

$$\begin{aligned} &V_{11} = S_{11}, \quad V_{21} = S_{21}, \qquad V_{22} = I_2, \\ &U_{11} = I_1, \quad\ \ U_{12} = S_{11}^{-1}S_{12}, \quad U_{22} = S_{22} - S_{21}S_{11}^{-1}S_{12}. \end{aligned} \tag{2.6}$$

(Obviously, here I_k denotes the identity operator in $\mathfrak{H}_k$.)

If relations (2.6) hold, the operator

$$V^{-1} = \begin{bmatrix} S_{11}^{-1} & 0 \\ -S_{21}S_{11}^{-1} & I_2 \end{bmatrix} \tag{2.7}$$

is bounded. In what follows we shall assume that $T = S^{-1}$ is bounded. Then the boundedness of V^{-1} and (2.4) imply the boundedness of

$$U^{-1} = \begin{bmatrix} U_{11}^{-1} & -U_{11}^{-1}U_{12}U_{22}-1 \\ 0 & U_{22}^{-1} \end{bmatrix}. \tag{2.8}$$

Notice that (2.7) and (2.8) imply

$$T_{22} = U_{22}^{-1}. \tag{2.9}$$

Denote by P_k $(k = 1, 2)$ the orthogonal projector from $\mathfrak{H}$ onto $\mathfrak{H}_k$.

Theorem 2.1 (Main factorization theorem [82]). *Suppose that the operator-valued function $W(z)$ and its inverse admit the realizations* (0.2) *and* (0.3), *respectively, and that there exists a bounded operator S satisfying* (2.1) *and* (2.2). *If S and S_{11} are invertible on $\mathfrak{H}$ and $\mathfrak{H}_1$, respectively, and if*

$$AP_2 = P_2AP_2, \qquad BP_1 = P_1BP_1, \tag{2.10}$$

then

$$W(z) = W_2(z)W_1(z), \tag{2.11}$$

where

$$W_1(z) = I - \Pi_2^* S_{11}^{-1}(A_{11} - zI)^{-1}P_1\Pi_1, \tag{2.12}$$

$$W_2(z) = I - \Gamma_2^*(A_{22} - zI)^{-1}T_{22}^{-1}P_2\Gamma_1. \tag{2.13}$$

Proof. First, we remark that from (2.9) it follows that T_{22} is invertible on $\mathfrak{H}_2$. In view of (0.2) and (2.2), we can write

$$W(z) = I - \Pi_2^* T(A - zI)^{-1}\Pi_1.$$

Hence, by Lemma 2.1,

$$W(z) = I - \Pi_2^* U^{-1}(\widetilde{A} - zI)^{-1}V^{-1}\Pi_1, \qquad \widetilde{A} = V^{-1}AV. \tag{2.14}$$

By (2.5), (2.10), and (2.14),

$$\widetilde{A}_{11} = V_{11}^{-1}A_{11}V_{11}, \qquad \widetilde{A}_{22} = V_{22}^{-1}A_{22}V_{22}, \qquad \widetilde{A}_{12} = 0. \tag{2.15}$$

Next, (2.1) and (2.4) yield

$$P_2 V^{-1}\Pi_1\Pi_2^* U^{-1}P_1 = P_2(V^{-1}AV - UBU^{-1})P_1. \tag{2.16}$$

By (2.10), we have $P_2BP_1 = 0$. Then it follows from (2.16) that

$$\widetilde{A}_{21} = P_2V^{-1}\Pi_1\Pi_2^*U^{-1}P_1. \tag{2.17}$$

Formula (2.11) follows directly from (2.14) and (2.17) upon setting

$$W_1(z) = I - \Pi_2^*U^{-1}(\widetilde{A}_{11} - zI)^{-1}P_1V^{-1}\Pi_1, \tag{2.18}$$

$$W_2(z) = I - \Pi_2^*U^{-1}(\widetilde{A}_{22} - zI)^{-1}P_2V^{-1}\Pi_1\,. \tag{2.19}$$

Using (2.6) and (2.15), one can verify that (2.18) and (2.19) are equivalent to (2.12) and (2.13), respectively. This completes the proof of the theorem. □

1.2.2. Theorem 2.1 is close in spirit to well-known theorems on the factorization of characteristics functions of operators obtained earlier by Livšic [58] and Brodskiĭ [14]. To elucidate the connection between these theorems, let us write the operator $A - A^*$ in the form

$$A - A^* = i\Pi_1 J\Pi_1^*, \qquad J = J^*, \qquad J^2 = I, \qquad J \in \{\mathfrak{G}, \mathfrak{G}\}. \tag{2.20}$$

Relation (2.20) is the special case of (2.1) when

$$S = I, \qquad B = A^*, \qquad \Pi_2^* = iJ\Pi_1^*. \tag{2.21}$$

In this case the transfer function $W_A(z)$, defined by (0.2), has the form

$$W_A(z) = I - iJ\Pi_1^*(A - zI)^{-1}\Pi_1 \tag{2.22}$$

and coincides with the characteristic function of the operator A [58]. Under the conditions (2.21), the well-known theorem on factorization of characteristic functions [58], [14] mentioned before follows from Theorem 2.1. A version of Theorem 2.1 closely related to the definition of the characteristic function of a quasi-unitary operator can be found in [99].

A result close to Theorem 2.1 was obtained in the book [7] by a geometric approach. The same book and the articles [8], [9] contain important applications of the factorization theorem.

1.2.3. The following assertions hold true.

Corollary 2.1. *Suppose that the conditions of Theorem* 2.1 *are satisfied and that* $\dim \mathfrak{G} < \infty$, $\dim \mathfrak{H} < \infty$. *If the realization* (0.2) *of* $W(z)$ *is minimal, then the realizations of* $W_1(z)$ *and* $W_2(z)$ *given by* (2.12) *and* (2.13), *respectively, are also minimal.*

Corollary 2.2. *Suppose that the conditions of Theorem* 2.1 *are satisfied. Then*

$$W_1^{-1}(z) = I + \Pi_2^* P_1 (B_{11} - zI)^{-1} S_{11}^{-1} P_1 \Pi_1, \tag{2.23}$$

$$W_2^{-1}(z) = I + \Gamma_2^* P_2 T_{22}^{-1} (B_{22} - zI)^{-1} P_2 \Gamma_1. \tag{2.24}$$

Indeed, from (2.1) and (2.10) we obtain

$$A_{11} S_{11} - S_{11} B_{11} = P_1 \Pi_1 \Pi_2^* P_1. \tag{2.25}$$

Using this relation, one can verify that (2.23) holds by multiplying the right-hand sides of (2.12) and (2.23). Equality (2.24) is established in a similar manner.

1.2.4. Let us consider separately the conditions for the solvability of the systems (2.1), (2.2) for the operator S.

Theorem 2.2. *Suppose that the realizations* (0.2), (0.3) *are minimal and that* $\dim \mathfrak{G} < \infty$, $\dim \mathfrak{H} < \infty$. *Then there exists a bounded operator S with bounded inverse satisfying* (2.1) *and* (2.2).

Proof. Set $B_1 = A - \Pi_1 \Gamma_2^*$. By Theorem 1.1,

$$W^{-1}(z) = I + \Gamma_2^* (B_1 - zI)^{-1} \Pi_1, \tag{2.26}$$

the realization (2.26) being minimal. Therefore, (0.3) and (2.26) yield minimal realizations of the same operator-valued function $W^{-1}(z)$.

Theorem 1.3 implies the existence of a bounded operator S with bounded inverse such that

$$B_1 = SBS^{-1}, \qquad S\Gamma_1 = \Pi_1, \qquad \Gamma_2^* S = \Pi_2^*,$$

as needed.

If S satisfies (2.1), then

$$S(B - zI)^{-1} - (A - zI)^{-1} S = (A - zI)^{-1} \Pi_1 \Pi_2^* (B - zI)^{-1}. \tag{2.27}$$

Multiplying both sides of (2.27) by Γ_2^* to the left and by Γ_1 to the right and using relations (0.2) and (0.3), we have

$$(\Gamma_2^* S - \Pi_2^*)(B - zI)^{-1} \Gamma_1 = \Gamma_2^* (A - zI)^{-1} (S\Gamma_1 - \Pi_1). \tag{2.28}$$

Similarly, one can verify that

$$\Pi_2^* (B - zI)^{-1} (T\Pi_1 - \Gamma_1) = (\Pi_2^* T - \Gamma_2^*)(A - zI)^{-1} \Pi_1, \tag{2.29}$$

where T is a solution of the equation

$$TA - BT = \Gamma_1 \Gamma_2^*. \tag{2.30}$$

□

Theorem 2.3. *Suppose that the realizations* (0.2) *and* (0.3) *are minimal. Then the following assertions are valid:*

I. *If S satisfies* (2.30) *and one of the conditions* (2.2), *then it also satisfies the other condition.*

II. *If T satisfies* (2.30) *and one of the conditions*

$$T\Pi_1 = \Gamma_1, \qquad \Pi_2^* T = \Gamma_2^*, \tag{2.31}$$

then it also satisfies the other condition.

III. *If S and T satisfy* (2.1), (2.2) *and* (2.30), (2.31), *respectively, then they are inverses to each other.*

IV. *If the spectra of A and B are disjoint, then there exists a bounded operator S with bounded inverse satisfying* (2.1), (2.2).

Proof. Assertions I and II are direct consequences of (2.28) and (2.29). To prove III, multiply to the left both sides of (2.1) by T: $TAS - TSB = \Gamma_1\Pi_2^*$. Hence, by (2.1) and (2.30),

$$BTS - TSB = 0. \tag{2.32}$$

From (2.2) and (2.31) we also obtain $TS\Gamma_1 = \Gamma_1$. Then we deduce from (2.32) that $TSB^n\Gamma_1 = B^n\Gamma_1$ $(n = 0, 1, 2, \dots)$. From the minimality of the realization (0.3) we conclude that $TS = I$. Similarly, $ST = I$, whence $T = S^{-1}$.

Finally, let us turn to assertion IV. Since the spectra of A and B are disjoint, equations (2.1) and (2.30) have solutions S and T. The equality (2.28) is possible only when its sides are simultaneously equal to zero. Then it follows from the minimality of (0.2) that (2.2) is satisfied. In much the same manner it can be shown that T satisfies (2.31). Hence, by assertion III, $T = S^{-1}$. The theorem is proved. □

1.3. Factorization of rational operator-valued functions

1.3.1. Let $W(z)$ be an operator-valued function of the form

$$W(z) = E + \sum_{k=1}^{m}\sum_{i=1}^{m_k} \frac{D_{ki}}{(z_k - z)^i}, \tag{3.1}$$

where $D_{ki} \in \{\mathfrak{G}, \mathfrak{G}\}$, $\dim \mathfrak{G} < \infty$, and $z_k \neq z_i$ for $k \neq i$. Then $W^{-1}(z)$ admits a similar representation:

$$W^{-1}(z) = I + \sum_{k=1}^{l}\sum_{i=1}^{l_k} \frac{E_{ki}}{(\zeta_k - z)^i}, \tag{3.2}$$

where $E_{ki} \in \{\mathfrak{G}, \mathfrak{G}\}$ and $\zeta_k \neq \zeta_j$ for $k \neq j$.

Suppose that each of the sets of numbers z_k $(1 \le k \le m)$ and ζ_k $(1 \le k \le l)$ is partitioned into two subsets, L_1, L_2 and M_1, M_2, respectively. We shall say that $W(z)$ admits a *factorization with respect to the partitions* L_1, L_2 and M_1, M_2 if

$$W(z) = W_2(z)W_1(z)$$

where $W_1(z)$ and $W_2(z)$ are rational operator-valued functions of the form (3.1) and $z_k \in L_i$ in representation of W_i while $\zeta_k \in M_i$ in the representation of $W_i^{-1}(z)$ for $i = 1, 2$.

1.3.2. Consider again a factorization of the rational operator-valued function $W(z)$ of the form (3.2). The operator-valued functions $W(z)$ and $W^{-1}(z)$ admit minimal realizations (see Section 1, Example 4), and

$$\dim \mathfrak{H} = \delta(W) < \infty.$$

The number $\delta(W)$ is called the *McMillan degree* of $W(z)$.

The operators A and B in the minimal realizations of $W(z)$ and $W^{-1}(z)$ can be replaced by similar linear operators. Hence, without loss of generality we can assume that A and B are given in the N-dimensional space of column vectors by block-matrices of the form

$$A = \begin{bmatrix} A_{11} & 0 \\ A_{21} & A_{22} \end{bmatrix} \}N_1 \text{ rows} \quad , \quad B = \begin{bmatrix} B_{11} & B_{12} \\ 0 & B_{22} \end{bmatrix} \}N_2 \text{ rows} \quad ,$$

where the spectra of A_{kk} and B_{kk} belong to L_k and M_k respectively $(k = 1, 2)$. By Theorem 2.2, there exists a bounded operator S with bounded inverse satisfying (2.1) and (2.3). Next, let us introduce the projection

$$P_1 = \begin{bmatrix} I & 0 \\ 0 & 0 \end{bmatrix} \}N_1 \text{ rows} \quad .$$

Now on the subspace $\mathfrak{H}_1 = P_1\mathfrak{H}$ define the operator $S_{11} = P_1 S P_1$.

Suppose we are given a factorization (2.11) with respect to the partitions L_1, L_2 and M_1, M_2. Consider minimal realizations of the factors

$$W_k(z) = I - \Gamma_2^{*(k)}(\widetilde{A}_k - zI)^{-1}\Pi_1^{(k)}, \quad k = 1, 2, \tag{3.3}$$

where

$$\widetilde{A}_k \in \{\mathfrak{H}_k, \mathfrak{H}_k\}, \qquad \Pi_1^{(k)}, \Gamma_2^{(k)} \in \{\mathfrak{G}, \mathfrak{H}_k\}.$$

The factorization (2.11) is said to be *minimal* if

$$\delta(W) = \delta(W_1) + \delta(W_2). \tag{3.4}$$

We remark that one always has

$$\delta(W) \le \delta(W_1) + \delta(W_2). \tag{3.5}$$

Theorem 3.1. *Let $W(z)$ be an operator-valued function of the form* (3.1). *For $W(z)$ to admit a minimal factorization with respect to the partitions L_1, L_2 and M_1, M_2 it is necessary and sufficient that the following conditions be satisfied:*

1) *$N_1 = N_2$;*
2) *The operator S_{11} is invertible on $\mathfrak{H}_1$.*

Proof. The sufficiency of conditions 1) and 2) follows from Theorem 2.1.

Let us establish the necessity of these conditions. By (3.3),

$$W_k^{-1}(z) = I + \Gamma_2^{*(k)}(\widetilde{B}_k - zI)^{-1}\Pi_1^{(k)}, \qquad k = 1, 2, \tag{3.6}$$

with

$$\widetilde{B}_k = \widetilde{A}_k - \Pi_1^{(k)}\Gamma_2^{(k)}.$$

Set

$$\mathfrak{H} = \mathfrak{H}_1 \oplus \mathfrak{H}_2, \qquad \widetilde{\Gamma}_2 = \widetilde{\Gamma}_2^{(1)} + \Gamma_2^{(2)}, \qquad \widetilde{\Pi}_1 = \Pi_1^{(1)} + \Pi_1^{(2)}.$$

Then Theorems 1.1 and 1.2. yield the relations

$$W(z) = I - \widetilde{\Gamma}_2^*(\widetilde{A} - zI)^{-1}\widetilde{\Pi}_1, \qquad W^{-1}(z) = I + \widetilde{\Gamma}_2^*(\widetilde{B} - zI)^{-1}\widetilde{\Pi}_1, \tag{3.7}$$

where

$$\widetilde{A} = \begin{bmatrix} \widetilde{A}_1 & 0 \\ \widetilde{A}_{21} & \widetilde{A}_2 \end{bmatrix}, \qquad \widetilde{A}_{21} = \Pi_1^{(2)}\Gamma_2^{(1)*}, \qquad \widetilde{B} = \widetilde{A} - \widetilde{\Pi}_1\widetilde{\Gamma}_2^* = \begin{bmatrix} \widetilde{A}_1 & \widetilde{B}_{12} \\ 0 & \widetilde{B}_2 \end{bmatrix}. \tag{3.8}$$

Realizations (3.7) are minimal because realizations (3.3), (3.6) are minimal and condition (3.4) is satisfied. Thus, $W(z)$ and $W^{-1}(z)$ admit two minimal realizations: (0.2), (0.3) and (3.7). By Theorem 1.3, there exist two invertible operators $U, V \in \{\mathfrak{H}, \widetilde{\mathfrak{H}}\}$ such that

$$UAU^{-1} = \widetilde{A}, \qquad \widetilde{\Gamma}_2^*U = \Gamma_2^*, \qquad U^{-1}\widetilde{\Pi}_1 = \Pi_1 \tag{3.9}$$

$$VBV^{-1} = \widetilde{A}, \qquad \widetilde{\Pi}_2^*V = \Pi_2^*, \qquad V^{-1}\widetilde{\Gamma}_1 = \Gamma_1. \tag{3.10}$$

Since A and $\widetilde{A}$ are similar, it follows that $\mathfrak{H} = \widetilde{\mathfrak{H}}$, $\mathfrak{H}_k = \widetilde{\mathfrak{H}}_k$, $k = 1, 2$. Therefore, by (3.8), we have $N_1 = N_2$. Comparing (2.1), (2.2) and (3.8)–(3.10), we conclude that

$$S = U^{-1}V. \tag{3.11}$$

Further, since A and $\widetilde{A}$, B and $\widetilde{B}$ have a block-triangular structure, so do U and V, that is

$$U = \begin{bmatrix} U_{11} & 0 \\ U_{21} & U_{22} \end{bmatrix}, \qquad V = \begin{bmatrix} V_{11} & V_{12} \\ 0 & V_{22} \end{bmatrix}. \tag{3.12}$$

Finally, since U and V are invertible, so are U_{11} and V_{11} on $\mathfrak{H}_1$. By (3.11) and (3.12), $S_{11} = U_{11}^{-1}V_{11}$. Hence, S_{11} is invertible on $\mathfrak{H}_1$. This completes the proof of the theorem. □

In Theorem 3.1 we followed N. M. Kostenko's arguments [44], adding the requirement that the factorization $W(z)$ be minimal. Without this minimality assumption the theorem is false. (Unfortunately, Kostenko's theorem was reproduced in our paper [84] without this requirement.) Here is an example of a rational matrix-valued function of the form (3.1) that does not have a minimal factorization.

Example 1. Let $W(z)$ have the form

$$W(z) = \begin{bmatrix} (\mu_2 - z)/(\lambda_1 - z) & 0 \\ 0 & (\mu_1 - z)/(\lambda_2 - z) \end{bmatrix}. \tag{3.13}$$

Assume that

$$\lambda_1 \neq \mu_2, \quad \lambda_2 \neq \mu_1. \tag{3.14}$$

Choosing

$$A = \begin{bmatrix} \lambda_1 & 0 \\ 0 & \lambda_2 \end{bmatrix}, \qquad B = \begin{bmatrix} \mu_1 & 0 \\ 0 & \mu_2 \end{bmatrix}, \qquad \Gamma_1 = \begin{bmatrix} 0 & 1 \\ 1 & 0 \end{bmatrix},$$

$$\Gamma_2^* = \begin{bmatrix} \lambda_1 - \mu_2 & 0 \\ 0 & \lambda_2 - \mu_1 \end{bmatrix}, \qquad \Pi_1 = \begin{bmatrix} 1 & 0 \\ 0 & 1 \end{bmatrix}, \qquad \Pi_2^* = \begin{bmatrix} 0 & \lambda_1 - \mu_2 \\ \lambda_2 - \mu_1 & 0 \end{bmatrix},$$

we obtain representations (0.2), (0.3) for $W(z)$ and $W^{-1}(z)$, respectively. It is readily verified that these realizations are minimal. By (2.1) and (2.2), the matrix S has the form

$$S = \begin{bmatrix} 0 & 1 \\ 1 & 0 \end{bmatrix}. \tag{3.15}$$

Let

$$\lambda_k \in L_k, \quad \mu_k \in M_k, \quad k = 1, 2. \tag{3.16}$$

Since $S_{11} = 0$, it follows that the matrix $W(z)$ defined by (3.13) does not admit a minimal factorization with respect to the partition (3.16).

Chapter 2
Operator Identities and S-nodes

In this chapter we continue the study of the operator identities that arose in connection with factorization problems in Chapter 1. These identities lead to the notion of an S-node, which is a generalization of the notion of node or colligation [58], [14] known in the theory of characteristic functions of operators. We shall focus on the operator identities that generate symmetric S-nodes. In this way we obtain new results and recover previously known results [65] on hereditary properties of factors arising in the factorization of an operator-valued function.

2.1. Elementary properties of S-nodes

2.1.1. Suppose that there are given operators

$$A, B, S \in \{\mathfrak{H}, \mathfrak{H}\}, \qquad \Pi_k, \Gamma_k \in \{\mathfrak{G}, \mathfrak{H}\} \qquad (k = 1, 2), \tag{1.1}$$

where $\mathfrak{G}$ and $\mathfrak{H}$ are Hilbert spaces. We say that the operators (1.1) form an S-*node* if the relations (2.1) and (2.2) of Chapter 1 hold. This notion of an S-node is a generalization of the earlier notion of a node [58], [14] and reduces to the latter when the relations (2.21) of Chapter 1 are satisfied as well. (The alternative terms *colligation* and S-*colligation* are often used.) As in Chapter 1, let us introduce the operator-valued functions

$$W_A(z) = I - \Gamma_2^*(A - zI)^{-1}\Pi_1, \qquad W_B(z) = I + \Pi_2^*(B - zI)^{-1}\Gamma_1. \tag{1.2}$$

Theorem 1.1. *Assume that z and ζ do not belong to the spectra of A and B, respectively. Then*

$$W_A(z)W_B(\zeta) = I - (z - \zeta)\Gamma_2^*(A - zI)^{-1}S(B - \zeta I)^{-1}\Gamma_1. \tag{1.3}$$

Proof. Indeed,

$$\begin{aligned} W_A(z)W_B(\zeta) = I - \Gamma_2^*(A - zI)^{-1}\Pi_1 + \Pi_2^*(B - \zeta I)^{-1}\Gamma_1 - \\ - \Gamma_2^*(A - zI)^{-1}\Pi_1\Pi_2^*(B - \zeta I)^{-1}\Gamma_1. \end{aligned} \tag{1.4}$$

It follows from relation (2.1) of Chapter 1 that

$$\Pi_1\Pi_2^* = (A - zI)S - S(B - \zeta I) + (z - \zeta)S. \tag{1.5}$$

Recalling relation (2.2) of Chapter 1, we see that (1.3) is a consequence of (1.4) and (1.5). □

For $z = \zeta$ (1.3) becomes

$$W_A(z)W_B(z) = I. \tag{1.6}$$

2.1.2. The operator-valued function $W_A(z)$ is invertible in a neighbourhood of $z = \infty$. Let $\mathcal{D}(A, B)$ be a connected set in the complex plane that contains the point $z = \infty$ but contains no points of the spectra of A and B. By (1.6),

$$W_B(z) = W_A^{-1}(z), \qquad z \in \mathcal{D}(A, B). \tag{1.7}$$

Corollary 1.1. *For $z \in \mathcal{D}(A, B)$ the equation*

$$SY(z) = (A - zI)^{-1}\Pi_1 \tag{1.8}$$

has the solution

$$Y(z) = (B - zI)^{-1}\Gamma_1 W_A(z). \tag{1.9}$$

Proof. By relation (2.1) of Chapter 1,

$$S(B - zI)^{-1} - (A - zI)^{-1}S = (A - zI)^{-1}\Pi_1\Pi_2^*(B - zI)^{-1}. \tag{1.10}$$

Therefore, relations (1.2) and (2.2) of Chapter 1 yield

$$S(B - zI)^{-1}\Gamma_1 = (A - zI)^{-1}\Pi_1 W_B(z), \tag{1.11}$$

from which the assertion of the corollary follows by multiplying both sides by $W_A(z)$ on the right and using (1.7). □

2.1.3 Let us examine separately a special case of S-node, when

$$\mathfrak{G} = \mathfrak{G}_1 \oplus \mathfrak{G}_1, \tag{1.12}$$

where $\mathfrak{G}_1$ is a Hilbert space ($\dim \mathfrak{G}_1 \leq \infty$). In this case the vectors $g \in \mathfrak{G}$ and the operators $\Pi_1, \Pi_2 \in \{\mathfrak{G}, \mathfrak{H}\}$ admit representations

$$g = \begin{bmatrix} g_1 \\ g_2 \end{bmatrix}, \qquad \Pi_1 = [\Phi_1 \;\; \Phi_2], \qquad \Pi_2 = [\Psi_1 \;\; \Psi_2], \tag{1.13}$$

where $g_k \in \mathfrak{G}_1$, $\Phi_k, \Psi_k \in \{\mathfrak{G}_1, \mathfrak{H}\}$, $k = 1, 2$. Relation (2.1) of Chapter 1 now takes the form

$$AS - SB = \Phi_1\Psi_1^* + \Phi_2\Psi_2^*. \tag{1.14}$$

Putting

$$\Gamma_1 = [N_1 \;\; N_2], \qquad \Gamma_2 = [M_1 \;\; M_2], \qquad N_k, M_k \in \{\mathfrak{G}_1, \mathfrak{H}\}, \quad k = 1, 2, \tag{1.15}$$

we rewrite relations (2.2) of Chapter 1 as

$$SN_k = \Phi_k, \qquad M_k^* S = \Psi_k^*, \qquad k = 1, 2. \tag{1.16}$$

By (1.12), the operator-valued functions $W_A(z)$ and $W_B(z)$ have the block decompositions

$$W_A(z) = \begin{bmatrix} a_1(z) & a_2(z) \\ b_1(z) & b_2(z) \end{bmatrix}, \qquad W_B(z) = \begin{bmatrix} c_1(z) & d_1(z) \\ c_2(z) & d_2(z) \end{bmatrix}. \tag{1.17}$$

Further, by Corollary 1.1, the equations

$$SY_k(z) = (A - zI)^{-1}\Phi_k, \qquad k = 1, 2, \tag{1.18}$$

have the following solutions:

$$Y_k(z) = (B - zI)^{-1}[N_1 a_k(z) + N_2 b_k(z)], \qquad k = 1, 2. \tag{1.19}$$

Now define the operator-valued function $\rho(z, \zeta)$ by

$$\rho(z, \zeta) = \Psi_1^*(B - zI)^{-1} Y_2(\zeta). \tag{1.20}$$

The reflection and transmission coefficients in radiative transfer problems and the scattering diagram in diffraction problems can be expressed in terms of $\rho(z, \zeta)$. The analytic structure of $\rho(z, \zeta)$ is characterized by the following theorem.

Theorem 1.2. *Suppose that relations* (1.14)–(1.16) *are satisfied and that* $z, \zeta \in \mathcal{D}(A, B)$. *Then*

$$\rho(z, \zeta) = \frac{c_1(z)a_2(\zeta) + d_1(z)b_2(\zeta)}{z - \zeta}. \tag{1.21}$$

Proof. Relations (1.19) and (1.20) yield

$$\rho(z, \zeta) = \frac{\Psi_1^*[(B - zI)^{-1} - (B - \zeta I)^{-1}][N_1 a_2(\zeta) + N_2 b_2(\zeta)]}{z - \zeta}.$$

Hence, by (1.2), (1.13), (1.15), and (1.17),

$$\rho(z, \zeta) = \frac{[c_1(z) - c_1(\zeta)]a_2(\zeta) + [d_1(z) - d_1(\zeta)]b_2(\zeta)}{z - \zeta}. \tag{1.22}$$

It follows from (1.7) and (1.17) that

$$c_1(z)a_2(z) + d_1(z)b_2(z) = 0. \tag{1.23}$$

Formula (1.21) is now seen to be a direct consequence of (1.22) and (1.23). □

2.1.4. Theorem 1.2 on the analytic structure of $\rho(z,\zeta)$ yields a substantial generalization of a well-known formula of Ambartsumyan [5] which concerns the case when

$$Sf = f(x) + \int_0^\omega f(t)k(x-t)\,dt, \qquad k(x) \in L(-\omega,\omega), \qquad k(x) = \overline{k(-x)}, \tag{1.24}$$

and

$$Af = i\int_0^x f(t)\,dt, \qquad B = A^*. \tag{1.25}$$

In [82] Ambartsumyan's formula was extended to other classes of integro-differential operators with kernels depending on the difference of the arguments. Here we consider a more general class of operators S of the form

$$Sf = \frac{d}{dx}\int_0^\omega s(x,t)f(t)\,dt, \tag{1.26}$$

where $f(t) = \operatorname{col}[f_1(t), f_2(t), \dots, f_n(t)]$ and the kernel $s(x,t)$ has the following structure:

$$s(x,t) = \{s_{l,m}(\omega_l x - \omega_m t)\}_{l,m=1}^n, \quad \omega_k = \bar{\omega}_k, \quad \omega_k \neq 0, \quad 1 \le k \le n. \tag{1.27}$$

We will assume that S is bounded in the space of vector-valued functions $L_n^2(0,\omega)$ and $s_{l,m}(v) \in L^2(-\omega_m\omega,\ \omega_l\omega)$. The adjoint S^* of the operator S acts as follows:

$$S^*f = \frac{d}{dx}\int_0^\omega s_1(x,t)f(t)dt, \tag{1.28}$$

where

$$s_1(x,t) = -W^{-1}s^*(t,x)W, \qquad W = \operatorname{diag}\{\omega_1, \omega_2, \dots, \omega_n\}. \tag{1.29}$$

Next, we take

$$Af = iW\int_0^x f(t)\,dt, \qquad M(x) = Ws(x,0), \qquad N(x) = -Ws(0,x). \tag{1.30}$$

A direct calculation yields the operator identity

$$(AS - SA^*)f = i\int_0^\omega [M(x) + N(t)]f(t)\,dt. \tag{1.31}$$

Let $\mathfrak{G}_1$ denote the space of n-dimensional column vectors g. Let us introduce the following operators acting from $\mathfrak{G}_1$ to $L_n^2(0,\omega)$:

$$\Phi_1 g = M(x)g, \qquad \Phi_2 g = g, \qquad \Psi_1 g = -ig, \qquad \Psi_2 g = -iN^*(x)g. \tag{1.32}$$

Then the identity (1.31) can be rewritten in the form (1.14), where $B = A^*$. Since

$$(A - zI)^{-1}\Phi_2 g = -\frac{1}{z}\exp\left(\frac{iWx}{z}\right)g, \tag{1.33}$$

Corollary 1.1 implies that in the case under consideration the following assertion holds true.

Corollary 1.2. *Let S be a bounded operator in $L^2_n(0,\omega)$ of the form (1.26), (1.27). If there exist matrix-valued functions $N_k(x)$, $M_k(x)$, $k = 1,2$ with columns in $L^2_n(0,\omega)$ and satisfying*

$$SN_1 = M, \qquad SN_2 = I_n, \qquad S^*M_1 = I_n, \qquad S^*M_2 = N^*, \tag{1.34}$$

then

$$SY_2(x,z) = -\frac{1}{z}\exp\left(\frac{iWx}{z}\right), \tag{1.35}$$

where

$$Y_2(x,z) = -\frac{1}{z}\left[u(x,z) - i\frac{W}{z}\int_x^\omega \exp\left(\frac{i(x-t)W}{z}\right)\cdot u(t,z)\,dt\right], \tag{1.36}$$

$$u(x,z) = N_1(x)a(z) + N_2(x)b(z), \tag{1.37}$$

$$a(z) = -\frac{i}{z}\int_0^\omega M_1^*(x)\exp\left(\frac{iWx}{z}\right)dx, \tag{1.38}$$

$$b(z) = -I_n - \frac{i}{z}\int_0^\omega M_2^*(x)\exp\left(\frac{iWx}{z}\right)dx. \tag{1.39}$$

Here we put $a_2(z) = a(z)$, $b_2(z) = b(z)$. Setting $c(z) = c_1(z)$, $d(z) = d_1(z)$, we deduce from (1.2) and (1.7) the expressions

$$c(z) = I_n - \frac{i}{z}\int_0^\omega \exp\left(\frac{-iWx}{z}\right)N_1(x)\,dx, \tag{1.40}$$

and

$$d(z) = -\frac{i}{z}\int_0^\omega \exp\left(\frac{-iWx}{z}\right)N_2(x)\,dx. \tag{1.41}$$

From (1.20), (1.21), (1.23) we obtain the following result.

Corollary 1.3. *Suppose that the conditions of Corollary 1.2 are satisfied. Then the operator-valued function*

$$\rho(z,\zeta) = -\frac{i}{z}\int_0^\omega \exp\left(\frac{iWx}{z}\right)Y_2(x,\zeta)\,dx \tag{1.42}$$

has the form

$$\rho(z,\zeta) = \frac{c(z)a(\zeta) + d(z)b(\zeta)}{z-\zeta}. \tag{1.43}$$

Let us note that for $\omega_1 = \omega_2 = \ldots = \omega_n$ the operator S of the form (1.26), (1.27) is an operator with a kernel depending on the difference of the arguments, that is, $s(x,t) = s(x-t)$. If $\omega_1{=}\omega_2{=}\ldots{=}\omega_p = a$, $\omega_{p+1}{=}\omega_{p+2}{=}\ldots{=}\omega_n{=}-a$ $(1 \le p < n)$, then $s(x,t)$ has the following structure

$$s(x,t) = \begin{bmatrix} s_{11}(x-t) & s_{12}(x+t) \\ s_{21}(x+t) & s_{22}(x-t) \end{bmatrix} \}p \text{ rows}.$$

2.2. Symmetric S-nodes

2.2.1. An S-node is said to be *symmetric* if

$$S = S^*, \qquad B = A^*, \qquad \Pi_2^* = iJ\Pi_1^*, \qquad \Gamma_2^* = iJ\Gamma_1^*, \tag{2.1}$$

where $J = J^* \in \{\mathfrak{G}, \mathfrak{G}\}$ and $J^2 = I$.

With a symmetric S-node we can associate the operator-valued transfer function $W_A(z)$. Then (0.2) can be recast as

$$W_A(z) = I - iJ\Gamma_1^*(A - zI)^{-1}\Pi_1. \tag{2.2}$$

Theorem 1.1 has the following immediate consequence.

Theorem 2.1 [82]. *In the case of symmetric S-nodes the reflection principle holds for $W_A(z)$, that is,*

$$W_A(z)\,J\,W_A^*(\bar{z}) = J, \qquad z \in \mathcal{D}(A, A^*). \tag{2.3}$$

The following assertion is in a certain sense the converse of Theorem 2.1.

Theorem 2.2. *Suppose that $W_A(z)$ admits a realization* (2.2) *and that $\Gamma_1\mathfrak{G}$ is a generating subspace for A^*, that is,*

$$\overline{\bigvee_{n=0}^{\infty} A^{*n}\Gamma_1\mathfrak{G}} = \mathfrak{H}. \tag{2.4}$$

Suppose further that the reflection principle (2.3) *holds and that there exists an operator S in $\{\mathfrak{H}, \mathfrak{H}\}$ such that*

$$AS - SA^* = i\Pi_1 J\Pi_1^*, \qquad S\Gamma_1 = \Pi_1. \tag{2.5}$$

Then $S = S^*$

Proof. By the reflection principle,

$$W_B(z) = JW_A^*(\bar{z})J = I + iJ\Pi_1^*(A^* - zI)^{-1}\Gamma_1. \tag{2.6}$$

Comparing (2.2) and (2.6) with (1.2), we obtain

$$B = A^*, \qquad \Pi_2^* = iJ\Pi_1^*, \qquad \Gamma_2^* = iJ\Gamma_1^*. \tag{2.7}$$

The relation $\Gamma_2^*S = \Pi_2^*$ follows from relation (2.28) of Chapter 1 and relations (2.4) and (2.5). Hence, relations (2.5) are equivalent to relations (2.1) and (2.2) of Chapter 1. Applying Theorem 1.1 and using (2.6), we obtain the relation

$$W_A(z)JW_A^*(\zeta) = J - i(z - \bar{\zeta})J\Gamma_1^*(A - zE)^{-1}S(A^* - \bar{\zeta}E)^{-1}\Gamma_1 J. \tag{2.8}$$

Now assume that the vectors h_j $(1 \le j \le m \le \infty)$ form a basis in $\mathfrak{G}$. For any natural number $n \le m$ and an arbitrary choice of points $z_k \in \mathcal{D}(A, A^*)$ such that $z_k \neq \bar{z}_l$, the quadratic form

$$R(\zeta_1, \zeta_2, \dots, \zeta_n) = i\sum_{k,l=1}^{n}\left(\frac{W_A(z_k)JW_A^*(z_l) - J}{z_k - \bar{z}_l}h_l, h_k\right)\zeta_l\bar{\zeta}_k \tag{2.9}$$

is real. Then by (2.4), (2.8), the form (Sf, f) is real for $f \in \mathfrak{H}$, that is, $S = S^*$. This completes the proof of the theorem. □

Corollary 2.1. *Any operator-valued function $W_A(z)$ that satisfies the conditions of Theorem 2.2 corresponds to a symmetric S-node A, S, Π_1, Γ_1.*

In fact, the operators B, Π_2, Γ_2 of the S-node are defined by the recipe (2.7).

2.2.2. An operator-valued function $W(z)$ is said to belong to the *class* $\mathcal{R}_\kappa$ if $W(z)$ is analytic in a neighbourhood D of $z = \infty$ and if for any choice of a positive integer n $(n \le m)$ and of complex points z_k $(z_k \ne \bar{z}_l,\ z_k \in D)$, the form (2.9) has at most κ negative squares, and for some choice of points z_k it has exactly κ negative squares. Classes of matrix-valued functions meromorphic in $\operatorname{Im} z > 0$ similar to the class $\mathcal{R}_\kappa$ were introduced by M. G. Kreĭn and H. Langer [51] in connection with interpolation problems.

A selfadjoint operator $S \in \{\mathfrak{H}, \mathfrak{H}\}$ is said to belong to the *class* $\mathcal{H}_\kappa$ if it has κ negative eigenvalues, counting multiplicities. From (2.4), (2.8), (2.9) we obtain the following result.

Corollary 2.2. *Suppose that $W_A(z)$ satisfies the conditions of Theorem 2.2. For $W_A(z)$ to belong to the class R_κ it is necessary and sufficient that $S \in \mathcal{H}_\kappa$.*

2.3. Inherited properties of factors

2.3.1. In some problems of mathematical and applied character the factorization of an operator-valued function $W(z)$ is of interest only if the factors $W_k(z)$ have certain properties inherited from $W(z)$. For a number of important problems this idea was systematically exploited in a series of works by V. P. Potapov (see [21], [65]). Theorem 2.1 of Chapter 1 gives a new approach to problems concerning inherited properties of factors. The point is that in some cases the construction of factors by means of formulas (2.12), (2.13) of Chapter 1 preserves certain properties of $W(z)$. To see this, consider an invertible operator $S \in \{\mathfrak{H}, \mathfrak{H}\}$, where $\mathfrak{H} = \mathfrak{H}_1 \oplus \mathfrak{H}_2$. Then the operators S and $T = S^{-1}$ admit block representations

$$S = \begin{bmatrix} S_{11} & S_{12} \\ S_{21} & S_{22} \end{bmatrix}, \qquad T = \begin{bmatrix} T_{11} & T_{12} \\ T_{21} & T_{22} \end{bmatrix}. \tag{3.1}$$

Lemma 3.1. *Let S and S_{11} be bounded operators with bounded inverse on $\mathfrak{H}$ and $\mathfrak{H}_1$, respectively. If $S \in \mathcal{H}_\kappa$, then $S_{11} \in \mathcal{H}_{\kappa_1}$ $T_{22} \in \mathcal{H}_{\kappa_2}$, with $\kappa = \kappa_1 + \kappa_2$.*

Proof. From relation (2.6) and (2.9) of Chapter 1 it follows that

$$T_{22}^{-1} = S_{22} - S_{21} S_{11}^{-1} S_{12}, \tag{3.2}$$

whence

$$S = \begin{bmatrix} S_{11} \\ S_{12} \end{bmatrix} S_{11}^{-1} [S_{11} \ \ S_{12}] + \begin{bmatrix} 0 & 0 \\ 0 & T_{22}^{-1} \end{bmatrix}. \tag{3.3}$$

Therefore, $\kappa_1 + \kappa_2 \geq \kappa$. In $\mathfrak{H}$ consider the vectors

$$h_p = \operatorname{col}[f_p, 0], \qquad h_{\kappa_1+r} = \operatorname{col}[-S_{11}^{-1}S_{12}g_r, g_r], \tag{3.4}$$

where $1 \leq p \leq \kappa_1$, $1 \leq r \leq \kappa_2$, and f_p and g_r are orthonormal systems of eigenvectors of the operators S_{11} and T_{22}^{-1} corresponding to negative eigenvalues λ_p and μ_r, respectively. A straightforward calculation shows that the matrix $\{(Sh_k, h_l)\}_{k,l=1}^{\kappa_1+\kappa_2}$ is diagonal and negative. Consequently, $\kappa_1 + \kappa_2 \leq \kappa$. The lemma is proved. □

Theorem 3.1. *Assume that the symmetric S-node satisfies the conditions of Theorem 2.1 of Chapter 1. Then, in the factorization of $W_A(z)$ via the formulas (2.11)–(2.13) of Chapter 1, the factors $W_1(z)$, $W_2(z)$ satisfy the reflection principle (2.3). If, in addition, (2.4) holds and $W_A(z) \in \mathcal{R}_\kappa$, then $W_1(z)$ and $W_2(z)$ belong to the classes to $\mathcal{R}_{\kappa_1}$ and $\mathcal{R}_{\kappa_2}$, respectively, where $\kappa_1 + \kappa_2 = \kappa$.*

Proof. By relations (2.1) and (2.10) of Chapter 1,

$$A_{11}S_{11} - S_{11}A_{11}^* = iP_1\Pi_1 J\Pi_1^* P_1. \tag{3.5}$$

Therefore, the node $A_{11}, S_{11}, \widetilde{\Pi}_1 = P_1\Pi_1$, $\widetilde{\Gamma}_1 = S_{11}^{-1}P_1\Pi_1$ is symmetric. This node is associated with the operator-valued function

$$W_1(z) = I - iJ\widetilde{\Gamma}_1^*(A_{11} - zI)^{-1}\widetilde{\Pi}_1, \tag{3.6}$$

which coincides with the first factor in the factorization of $W(z)$ (see (2.12), Chapter 1). The reflection principle for $W_1(z)$ follows from Theorem 2.1. The reflection principle for $W_2(z)$ is established in the same manner.

To prove the second assertion of the theorem, write

$$\begin{aligned} & i\sum_{k,l=1}^{n}\left(\frac{W_A(z_k)JW_A^*(z_l) - J}{z_k - \bar{z}_l}h_l, h_k\right)\zeta_l\bar{\zeta}_k = \\ &= i\sum_{k,l=1}^{n}\left(W_2(z_k)\frac{W_1(z_k)JW_1^*(z_l) - J}{z_k - \bar{z}_l}W_2^*(z_l)h_l, h_k\right)\zeta_l\bar{\zeta}_k + \\ &\quad + i\sum_{k,l=1}^{n}\left(\frac{W_2(z_k)JW_2^*(z_l) - J}{z_k - \bar{z}_l}h_l, h_k\right)\zeta_l\bar{\zeta}_k. \end{aligned} \tag{3.7}$$

This shows that if $W_1(z) \in \mathcal{R}_{\kappa'}$ and $W_2(z) \in \mathcal{R}_{\kappa''}$, then

$$\kappa' + \kappa'' \geq \kappa. \tag{3.8}$$

Using Lemma 2.1, we conclude that $S_{11} \in \mathcal{H}_{\kappa_1}$, $T_{22} \in \mathcal{H}_{\kappa_1}$ and

$$\kappa_1 + \kappa_2 = \kappa. \tag{3.9}$$

By (3.5) and (3.6), formula (2.8) takes the form

$$\frac{i[W_1(z)JW_1^*(\zeta) - J]}{z - \bar{\zeta}} = J\widetilde{\Gamma}_1^*(A_{11} - zI)^{-1}S_{11}(A_{11}^* - \bar{\zeta}I)^{-1}\widetilde{\Gamma}_1 J, \tag{3.10}$$

from which it readily follows that $\kappa' \leq \kappa_1$. Similarly, $\kappa'' \leq \kappa_2$. It now follows from (3.8) and (3.9) that $\kappa' = \kappa_1$, $\kappa'' = \kappa_2$ and $\kappa' + \kappa'' = \kappa$. This completes the proof of the theorem. □

From Theorem 3.2 of Chapter 1 and Theorem 3.1 we obtain the following assertion due to V. P. Potapov [65].

Corollary 3.1. *Let $W(z)$ be a rational operator-valued function of the form* (3.1) *of Chapter* 1 *belonging to the class $\mathcal{R}_0$. Then relation* (3.13) *of Chapter* 1 *holds, where the factors $W_k(z)$ have the form* (3.14) *of Chapter* 1 *and belong to $\mathcal{R}_0$.*

Remark 3.1. For a rational operator-valued function $W(z)$ membership in the class $\mathcal{R}_0$ is equivalent to the inequality

$$\frac{i[W(z)JW^*(z) - J]}{z - \bar{z}} \geq 0.$$

Chapter 3
Continual Factorization

Under certain conditions the factorization theorem allows us to find a continual factorization of an operator-valued function $W(z)$, that is, to represent $W(z)$ as a multiplicative integral

$$W(z) = \int_a^{\curvearrowleft b} \exp\left[-\frac{dB(t)}{\alpha(t) - z}\right]. \tag{0.1}$$

The multiplicative integral is defined as

$$\int_a^{\curvearrowleft b} \exp\left[-\frac{dB(t)}{\alpha(t) - z}\right] = \lim_{\max \Delta t_j \to 0} \exp\left[\frac{\Delta_n B(t)}{\alpha(t_n) - z}\right] \dots \exp\left[\frac{\Delta_1 B(t)}{\alpha(t_1) - z}\right], \tag{0.2}$$

where $a = t_0 < t_1 < t_2 < \ldots \leq t_n = b$, $\Delta_k B(t) = B(t_k) - B(t_{k-1})$, $\Delta t_k = t_k - t_{k-1}$.

In this chapter we describe a method for constructing $B(t)$ for a given operator-valued function $W(z)$. This result will be used in the following chapters to solve a wide range of inverse problems.

We will consider separately the important case when the boundedness condition

$$\|W(z)\| \leq M, \qquad z \neq \bar{z}, \tag{0.3}$$

is satisfied.

3.1. The main continual factorization theorem

3.1.1. Let $W(z)$ be an operator-valued function defined on a Hilbert space $\mathfrak{G}$ ($\dim \mathfrak{G} \leq \infty$) which admits the representations (0.2) and (0.3) of Chapter 1. Introduce the operator S defined by the relations (2.1), (2.2) of Chapter 1. Further, assume that there exists a continuous increasing family of orthogonal projectors P_ζ $(a \leq \zeta \leq b)$ in the space $\mathfrak{H}$ such that $P_a = 0$, $P_b = E$, and

$$BP_\zeta = P_\zeta BP_\zeta, \qquad AQ_\zeta = Q_\zeta AQ_\zeta, \qquad Q_\zeta = E - P_\zeta, \qquad a \leq \zeta \leq b. \tag{1.1}$$

Denote

$$\mathcal{P}_1 = P_\zeta, \qquad \mathcal{P}_2 = P_{\zeta+\Delta\zeta} - P_\zeta, \qquad T_{ij} = \mathcal{P}_i S^{-1}_{\zeta+\Delta\zeta} \mathcal{P}_j, \qquad A_{ij} = \mathcal{P}_i A \mathcal{P}_j. \tag{1.2}$$

The following statement is the main result of this section.

Theorem 1.1 (Continual factorization teorem). *Suppose that the following conditions are satisfied:*

I. *The operator $S_\zeta = P_\zeta S P_\zeta$ is invertible on the subspace $\mathfrak{H}_\zeta = P_\zeta \mathfrak{H}$, and for some constant $C > 0$*

$$\|S_\zeta^{-1}\| \le C, \qquad a \le \zeta \le b. \tag{1.3}$$

II. *The variation of the operator-valued function*

$$B(\zeta) = \Pi_2^* S_\zeta^{-1} P_\zeta \Pi_1 \tag{1.4}$$

is bounded in norm, i.e., for some constant $C_1 > 0$ and any partition $a = x_0 < x_1 < \ldots < x_n = b$ of the interval $[a, b]$ we have

$$\sum_{k=1}^{n} \|B(x_k) - B(x_{k-1})\| \le C_1. \tag{1.5}$$

III. *There exists a continuous function $\alpha(\zeta)$ such that*

$$\|\Pi_2^* S_{\zeta+\Delta\zeta}^{-1}\{(A_{22} - zI)^{-1} - [\alpha(z) - z]^{-1} I\} T_{22}^{-1} \mathcal{P}_2 S_{\zeta+\Delta\zeta}^{-1} \Pi_1\| = o(\Delta\zeta), \tag{1.6}$$

whenever z does not belong to the range of the function $\alpha(\zeta)$, $a \le \zeta \le b$.

Then the operator-valued function $W(z)$, defined by relations (0.2) and (0.3) of Chapter 1, admits a representation as a multiplicative integral

$$W(z) = \int_a^{b} \!\!\!\!\!\! ^{\curvearrowleft} \exp\left[-\frac{dB(t)}{\alpha(t) - z}\right]. \tag{1.7}$$

Proof. Set $S_{ij} = \mathcal{P}_i S_{\zeta+\Delta\zeta} \mathcal{P}_j$. Then the invertibility of S_{11} and $S_{\zeta+\Delta\zeta}$ follows from (1.3). By relations (2.6) and (2.9) of Chapter 1,

$$T_{22}^{-1} = S_{22} - S_{21} S_{11}^{-1} S_{12}, \tag{1.8}$$

that is, the operator T_{22} is invertible.

Consider the operator-valued functions

$$\Gamma_1(\zeta) = S_\zeta^{-1} P_\zeta \Pi_1, \qquad \Gamma_2^*(\zeta) = \Pi_2^* S_\zeta^{-1} P_\zeta, \tag{1.9}$$

$$W(\zeta, z) = I - \Pi_2^* S_\zeta^{-1} (A_{11} - zI)^{-1} P_\zeta \Pi_1. \tag{1.10}$$

Applying the factorization Theorem 2.1 of Chapter 1, we obtain

$$W(\zeta + \Delta\zeta, z) = V(\zeta, \Delta\zeta, z) W(\zeta, z), \tag{1.11}$$

where

$$V(\zeta,\Delta\zeta,z) = I - \Gamma_2^*(\zeta+\Delta\zeta)(A_{22}-zI)^{-1}T_{22}^{-1}\mathcal{P}_2\Gamma_1(\zeta+\Delta\zeta). \tag{1.12}$$

Now consider the difference

$$W(\zeta+\Delta\zeta,z) - W(\zeta,z) = [V(\zeta,\Delta\zeta,z)-E]W(\zeta,z). \tag{1.13}$$

Using condition (1.6), we deduce from (1.12) and (1.13) that

$$\begin{aligned} &W(\zeta+\Delta\zeta,z) - W(\zeta,z) \\ &= \left\{-\Gamma_2^*(\zeta+\Delta\zeta)T_{22}^{-1}\ \mathcal{P}_2\Gamma_1(\zeta+\Delta\zeta)[\alpha(\zeta)-z]^{-1} + o(\Delta\zeta)\right\} W(\zeta,z). \end{aligned} \tag{1.14}$$

Let us prove the equality

$$S_{\zeta+\Delta\zeta}^{-1}T_{22}^{-1}\mathcal{P}_2S_{\zeta+\Delta\zeta}^{-1}P_{\zeta+\Delta\zeta} = S_{\zeta+\Delta\zeta}^{-1}P_{\zeta+\Delta\zeta} - S_\zeta^{-1}P_\zeta. \tag{1.15}$$

To this end, we write the right-hand side of (1.15) in the form

$$\begin{aligned} &\begin{bmatrix} T_{11} & T_{12} \\ T_{21} & T_{22} \end{bmatrix} \begin{bmatrix} 0 & 0 \\ 0 & T_{22}^{-1} \end{bmatrix} \begin{bmatrix} T_{11} & T_{12} \\ T_{21} & T_{22} \end{bmatrix} = \\ &= \begin{bmatrix} T_{11} & T_{12} \\ T_{21} & T_{22} \end{bmatrix} - \begin{bmatrix} T_{11}-T_{12}T_{22}^{-1}T_{21} & 0 \\ 0 & 0 \end{bmatrix}. \end{aligned} \tag{1.16}$$

Since

$$S_\zeta^{-1}P_\zeta = \begin{bmatrix} T_{11}-T_{12}T_{22}^{-1}T_{21} & 0 \\ 0 & 0 \end{bmatrix},$$

relation (1.15) is a consequence of relation (1.16). By (1.4) and (1.9), the equality (1.14) can be rewritten in the form

$$W(\zeta+\Delta\zeta) - W(\zeta,z) = -\frac{1}{\alpha(\zeta)-z}\left[B(\zeta+\Delta\zeta) - B(\zeta) + o(\Delta\zeta)\right] W(\zeta,z). \tag{1.17}$$

It follows from condition II of the theorem and (1.17) that

$$W(\zeta,z) = \int_0^{\zeta}\!\!\!\!\!\curvearrowleft \exp\left[-\frac{dB(t)}{\alpha(t)-z}\right], \tag{1.18}$$

which upon setting $\zeta = b$ completes the proof of the theorem. □

3.1.2. For operator-valued functions $W(z)$ in the class $\mathcal{R}_0$ (see Chapter 2, Section 2) the following result holds true.

Theorem 1.2. *Suppose that the operator-valued function $W(z)$ in the class $\mathcal{R}_0$ satisfies the following conditions:*

I. *$W(z)$ admits a realization of the form (2.2), Chapter 2, such that the operator $\Pi_1^*\Pi_1$ is trace class and relation (2.4) of Chapter 2 holds.*

II. *There exists a bounded operator S with bounded inverse satisfying relation (2.5) of Chapter 2.*

III. *There exists a continuous increasing family of orthogonal projectors P_ζ ($a \leq \zeta \leq b$, $P_a = 0$, $P_b = E$), such that $A^*P_\zeta = P_\zeta A^* P_\zeta$ for all ζ.*

IV. *There exists a continuous function $\alpha(\zeta)$ such that*

$$\|\mathcal{P}_2[A - \alpha(\zeta)I]\mathcal{P}_2\| < M|\Delta\zeta|. \tag{1.19}$$

Then $W(z)$ admits a representation (1.7) with a continuous operator-valued function $B(\zeta)$ on $[a, b]$ whose variation is bounded in norm.

Proof. By Corollary 2.2 of Chapter 2, the operator S is positive. Hence, condition I of Theorem 1.1 is satisfied. By (1.4),

$$B(\zeta) = iJB_1(\zeta), \tag{1.20}$$

where $B_1(\zeta) = \Pi_1^* S_\zeta^{-1} P_\zeta \Pi_1$. Next, it follows from (1.15) that

$$B_1(\zeta + \Delta\zeta) - B_1(\zeta) = L^*(\zeta, \Delta\zeta)L(\zeta, \Delta\zeta) \geq 0, \qquad \Delta\zeta > 0, \tag{1.21}$$

where

$$L(\zeta, \Delta\zeta) = T_{22}^{-1/2}\mathcal{P}_2 S_{\zeta+\Delta\zeta}^{-1} P_{\zeta+\Delta\zeta}\Pi_1. \tag{1.22}$$

The fact that $\Pi_1^*\Pi_1$ is trace class implies that the operators $B_1(\zeta)$ are trace class. Further, inequality (1.21) shows that the function $\rho(\zeta) = \operatorname{Tr} B_1(\zeta)$ is monotonically increasing and one has the inequality

$$\|B_1(\zeta + \Delta\zeta) - B_1(\zeta)\| \leq \rho(\zeta + \Delta\zeta) - \rho(\zeta). \tag{1.23}$$

Hence, $B_1(\zeta)$ has bounded variation, that is, condition II of Theorem 1.1 is satisfied.

Now let us show that $\rho(\zeta)$ is continuous. By (1.20) and (1.22),

$$\rho(\zeta + \Delta\zeta) - \rho(\zeta) = \operatorname{Tr}[L^*(\zeta, \Delta\zeta)L(\zeta, \Delta\zeta)]. \tag{1.24}$$

Let $\{g_i\}_{i=1}^\infty$ be an orthonormal basis in $\mathcal{G}$ and set $f_i = \mathcal{P}_2 S_{\zeta+\Delta\zeta}^{-1} P_{\zeta+\Delta\zeta}\Pi_1 g_i$. It follows from (1.24) that, for some constant $C > 0$,

$$\rho(\zeta + \Delta\zeta) - \rho(\zeta) \leq C \sum_{j=1}^{\infty} \|f_j\|^2 \leq \widetilde{C} \sum_j \|\Pi_1 g_j\|^2 < \infty. \tag{1.25}$$

Since $\|f_j\| \to 0$ as $\Delta\zeta \to 0$, (1.25) shows that $\rho(\zeta)$ is continuous. Consequently, the operator-valued functions $B_1(z)$ and $B(z)$ are continuous and the following estimate holds uniformly in ζ (see (1.24)):

$$\|L(\zeta, \Delta\zeta)\| = o(1), \qquad \Delta\zeta \to 0. \tag{1.26}$$

It follows from (1.19) and (1.26) that condition III of Theorem 1.1 is satisfied. Thus, all the conditions of Theorem 1.1 are satisfied. Consequently, (1.7) holds, which completes the proof of Theorem 1.2. □

Results similar to Theorem 1.2 were obtained earlier, in the construction multiplicative representations of characteristic matrix functions [32].

Remark 1.1. In the space $L^2_n(a,b)$ of n-dimensional vector-valued functions $f(t) = \mathrm{col}[f_1(t), f_2(t), \ldots, f_n(t)]$ consider the family of orthogonal projectors

$$(P_\zeta f)(x) = \begin{cases} f(x), & \text{for } \leq x \leq \zeta, \\ 0, & \text{for } \zeta < x \leq b. \end{cases} \tag{1.27}$$

Then for the choice

$$Af = iD\int_0^x f(t)\,dt, \qquad \alpha(\zeta) = 0, \tag{1.27}$$

$D = \mathrm{diag}\{d_1, d_2, \ldots, d_n\}$, and also for

$$Af = xf(x), \tag{1.28}$$

conditions III and IV of Theorem 1.2 hold.

Remark 1.2. If we assume, in addition, that the operator S admits a factorization [22] with respect to the chain P_ζ, then Theorem 1.2 can be deduced from well-known results found in [9], [45].

3.2. Bounded operator-valued functions

In this section the continual factorization theorem given in Section 1 is applied to the case when operator-valued function $W(z)$ of class $\mathcal{R}_0$ (see Chapter 1, Section 2) is analytic in the domain $z \neq \bar{z}$ and satisfies the inequality

$$\|W(z)\| \leq M, \qquad z \neq \bar{z} \tag{2.1}$$

for some constant $M > 0$. The resulting continual factorization of $W(z)$ can be interpreted as a solution of a certain inverse problem. For the case $J = E$ this makes contact with a well-known result of P. Masani and N. Wiener. Since $W(\cdot) \in \mathcal{R}_0$, $z = \infty$ is a regular point of $W(z)$. We will use the *normalization condition*

$$\text{s} - \lim_{z \to \infty} W(z) = I, \tag{2.2}$$

where the limit is taken in the strong sense. By the generalized Fatou theorem, the estimate (2.1) guarantees the almost everywhere existence of the limits [79]

$$W_{\pm}(x) = \text{s} - \lim_{y \to \pm 0} W(z), \qquad z = x + iy. \tag{2.3}$$

Using relations (2.1)–(2.3), we can write

$$W(z) = I + \frac{1}{2\pi i} \int_a^b \frac{W_+(x) - W_-(x)}{x - z}\, dx, \tag{2.4}$$

where the integral converges in the weak sense and $-\infty < a < b < \infty$. Since the class of operator-valued functions under consideration obeys the reflection principle

$$W^{-1}(z) = JW^*(\bar{z})J, \tag{2.5}$$

we conclude from (2.1) that $W^{-1}(z)$ is bounded. Consequently,

$$W^{-1}(z) = I - \frac{1}{2\pi i} \int_a^b \frac{W_+(x)^{-1} - W_-(x)^{-1}}{x - z}\, dx, \qquad z \neq \bar{z}. \tag{2.6}$$

3.2.2. Consider the polar decomposition ([65], Ch. 1)

$$W_+(x) = U(x)R(x), \tag{2.7}$$

where the operator-valued functions $U(x)$ and $R(x)$ are such that

$$U^*(x)JU(x) = J, \qquad JR(x) = R^*(x)J, \tag{2.8}$$

and in addition the spectrum of $R(x)$ is positive.

Using relation (2.5), we deduce from (2.7) and (2.8) that

$$W_-(x) = U(x)R^{-1}(x). \tag{2.9}$$

By relations (2.7) and (2.9),

$$R^2(x) = JW_+^*(x)JW_+(x), \qquad R^{-2}(x) = JW_-^*(x)JW_-(x), \tag{2.10}$$

which permits us to express $R(x)$ in terms of $W_{\pm}(x)$.

Lemma 2.1. *The operator-valued functions $R(x)$, $R^{-1}(x)$, $U(x)$, $U^{-1}(x)$ are bounded on $[a, b]$.*

Proof. According to the theory of J-moduli ([65], Ch. 1)

$$0 \leq J - JR^{-1}(x) \leq J - JR^{-2}(x), \qquad 0 \leq JR(x) - J \leq JR^2(x) - J. \tag{2.11}$$

It follows from (2.1) and relations (2.10) that the operator-valued functions $R^2(x)$ and $R^{-2}(x)$ are bounded on $[a, b]$. Then thanks to (2.11) the operator-valued functions $R(x)$ and $R^{-1}(x)$ are bounded on $[a, b]$. Formulas (2.1), (2.7) and (2.9) then imply that the operator-valued functions $U(x)$ and $U^{-1}(x)$ are bounded on $[a, b]$. The lemma is proved. □

It follows from (2.11) that

$$\mathcal{D}(x) = J\left[R(x) - R^{-1}(x)\right] \geq 0. \tag{2.12}$$

Let us introduce the following notation:

$$F_1(x) = \mathcal{D}^{1/2}(x), \qquad F_2(x) = \mathcal{D}^{1/2}(x)JU^*(x), \tag{2.13}$$

$$L(x) = \left\{I + \frac{1}{4}[F_1(x)JF_1(x)]^2\right\}^{1/2}. \tag{2.14}$$

Notice that, by Lemma 2.1, the operator-valued functions $\mathcal{D}(x)$, $F_1(x)$, $F_2(x)$, and $L(x)$ are bounded on $[a, b]$. Together with the space $\mathfrak{G}$ in which the operator-valued function $W(z)$ acts, we consider the space $\mathfrak{H}$ consisting of $\mathfrak{G}$-valued functions $f(x)$, equipped with the inner product

$$(f_1(x), f_2(x))_{\mathfrak{H}} = \int_a^b (f_1(x), f_2(x))_{\mathfrak{G}}\, dx. \tag{2.15}$$

3.2.3. Let us define the operators Π_1 and Γ_2, belonging to $\{\mathfrak{G}, \mathfrak{H}\}$ by the formulas

$$\Pi_1 g = \frac{1}{\sqrt{2\pi}} F_1(x)g, \qquad \Gamma_2 g = -\frac{i}{\sqrt{2\pi}} F_2(x)g, \qquad g \in \mathfrak{G}. \tag{2.16}$$

Then

$$\Pi_1^* f(x) = \frac{1}{\sqrt{2\pi}} \int_a^b F_1(x)f(x)\, dx, \qquad \Gamma_2^* f(x) = \frac{i}{\sqrt{2\pi}} \int_a^b F_2^*(x)f(x)\, dx. \tag{2.17}$$

By (2.7), (2.9), and (2.12), (2.13), we have

$$W_+(x) - W_-(x) = F_2^*(x)F_1(x), \tag{2.18}$$

and relations (2.4), (2.6) can be rewritten as realizations

$$W(z) = I - \Gamma_2^*(A - zI)^{-1}\Pi_1, \qquad W^{-1}(z) = I + \Pi_2^*(A - zI)^{-1}\Gamma_1, \tag{2.19}$$

where

$$\Pi_2 = -i\Pi_1 J, \qquad \Gamma_1 = i\Gamma_2 J \tag{2.20}$$

and the operator A acts according to the rule

$$Af = xf(x), \qquad f(x) \in \mathfrak{H}. \tag{2.21}$$

Next, consider the operators

$$Sf = L(x)f(x) + \frac{i}{2\pi}\text{P.V.} \int_a^b \frac{F_1(x)JF_1(t)}{x - t} f(t)\, dt, \tag{2.22}$$

$$Tf = L(x)f(x) - \frac{i}{2\pi}\text{P.V.} \int_a^b \frac{F_2(x)JF_2^*(t)}{x - t} f(t)\, dt, \tag{2.23}$$

where the symbol P.V. indicates that the integral is understood as principal value. The fact that the operator V defined as

$$Vf = \frac{1}{\pi}\text{P.V.}\int_{-\infty}^{\infty} \frac{f(t)}{x-t}\,dt,$$

is unitary and the operator-valued functions $L(x), F_1(x), F_2(x)$ are bounded implies that the operators S and T are bounded. A straightforward calculation shows that

$$AS - SA = i\Pi_1 J\Pi_1^*, \qquad TA - AT = i\Gamma_1 J\Gamma_1^*. \tag{2.24}$$

By (2.13) and (2.14), $L^2(x)F_1(x) = F_1(x)[R(x)+R^{-1}(x)]^2/4$, which in view of the fact that the spectra of $L(x)$ and $[R(x)+R^{-1}(x)]/2$ are positive implies that

$$L(x)F_1(x) = F_1(x)[R(x)+R^{-1}(x)]/2. \tag{2.25}$$

Using the familiar equality

$$W_+(x) + W_-(x) = 2I + \frac{i}{\pi}\text{P.V.}\int_a^b \frac{W_+(t)-W_-(t)}{x-t}\,dt,$$

we deduce fom relations (2.13), (2.14), and (2.25) that

$$S\Gamma_1 = \Pi_1, \qquad T\Pi_1 = \Gamma_1. \tag{2.26}$$

Denote by $\mathfrak{H}_1 \subset \mathfrak{H}$ the closure of the set of all vectors of the form $g(x) = F_1(x)f(x)$, where $f(x) \in \mathfrak{H}$. Also, let $\mathfrak{H}_2 = \mathfrak{H} \ominus \mathfrak{H}_1$. The subspaces $\mathfrak{H}_1$ and $\mathfrak{H}_2$ are invariant under the operators A, S and T. Indeed, it follows from formula (2.22) that $Sf - L(x)f \in \mathfrak{H}_1$ for all $f(x) \in \mathfrak{H}$. Since $L(x)F_1(x) = F_1(x)L(x)$ we have that $Sf \in \mathfrak{H}_1$ whenever $f \in \mathfrak{H}_1$, i.e., $\mathfrak{H}_1$ is an invariant subspace of the operator S. In the same manner it can be proved that $\mathfrak{H}_1$ is an invariant subspace of the operator T. The proof uses the fact that, by (2.13), $F_2(x) = F_1(x)JU^*(x)$. Further, $\mathfrak{H}_1$ is invariant under A because $xF_1(x) = F_1(x)x$. Finally, the subspace $\mathfrak{H}_2 = \mathfrak{H} \ominus \mathfrak{H}_1$ is invariant under the operators A, S and T because these operators are selfadjoint.

Note that the subspace $\mathfrak{H}_2$ is defined by the relation

$$F_1(x)f(x) \equiv 0, \qquad f(x) \in \mathfrak{H}_2.$$

Considering the realizations (2.19) on $\mathfrak{H}_1$, we see that, thanks to relations (2.24) and (2.26), the conditions of Theorem 2.2 of Chapter 2 are satisfied. Then, according to Corollaries 2.2 and 2.4 of Chapter 2, the operators S and T are positive and inverse to one another on $\mathfrak{H}_1$. Since S and T coincide with the identity operator on $\mathfrak{H}_2$, we conclude that S and T are positive and inverse to one another on the whole space $\mathfrak{H}$. Therefore, the following operator is positive and invertible:

$$S_\zeta f = L(x)f(x) + \frac{i}{2\pi}\text{P.V.}\int_a^\zeta \frac{F_1(x)JF_1(t)}{x-t}f(t)\,dt, \qquad a \le x \le \zeta. \tag{2.27}$$

Set

$$\Phi(\zeta, x) = S_\zeta^{-1} F_1(x). \tag{2.28}$$

Then, by (1.4),

$$B(\zeta) = iJ \frac{1}{2\pi} \int_a^\zeta F_1(x)\Phi(\zeta, x)\, dx = iJB_1(\zeta). \tag{2.29}$$

The following assertion is a direct consequence of Theorem 1.2.

Theorem 2.1. *Suppose that the operator-valued function $W(z)$ of class $\mathcal{R}_0$ satisfies condition* (2.1). *If the operator $\Pi_1^*\Pi_1$ is trace class, then $W(z)$ admits a multiplicative representation*

$$W(z) = \int_a^{\overset{\curvearrowleft}{b}} \exp\left[-\frac{dB(\zeta)}{\zeta - z}\right], \tag{2.30}$$

where $B(\zeta)$ is defined by (2.13), (2.14), *and* (2.27)–(2.29).

Remark 2.1. The operator $\Pi_1^*\Pi_1$ is of the form

$$\Pi_1^*\Pi_1 = \frac{1}{2\pi} \int_a^b \mathcal{D}(t)\, dt. \tag{2.31}$$

3.2.4. Let us introduce the operator-valued function

$$W(x, z) = \int_a^{\overset{\curvearrowleft}{b}} \exp\left[-iJ\frac{dB_1(\zeta)}{\zeta - z}\right], \qquad a \le x \le b, \tag{2.32}$$

where $B_1(\zeta)$ is a continuous, monotonically increasing operator-valued function of bounded variation. It is readily seen that $W(x, z)$ satisfies the system of equations

$$W(x, z) = I - iJ \int_a^x \frac{dB_1(\zeta)}{\zeta - z} W(\zeta, z), \qquad a \le x \le b. \tag{2.33}$$

The operator-valued function $W(b, z)$ is called the *Wronskian* of the system of equations (2.33).

Remark 2.2. Let $\rho(x) = \operatorname{Tr} B_1(x)$ be an absolutely continuous function and assume that $\dim \mathfrak{G} < \infty$. Then the system (2.33) is equivalent to the system of differential equations

$$\frac{dW(x, z)}{dx} = -iJ\frac{B_1'(x)W(x, z)}{x - z}, \qquad W(0, z) = I. \tag{2.34}$$

Theorem 2.1 gives a procedure for recovering systems of the form (2.33) and (2.34) from their Wronskian, that is, it can be restated as follows in the terms of the inverse problem.

Theorem 2.2. *Suppose that an operator-valued function* $W(z)$ *satisfies the conditions of Theorem* 2.1. *Then* $W(z)$ *is the Wronskian of a system of form* (2.33), *where the operator-valued function* $B_1(z)$ *is constructed by means of the formulas* (2.13), (2.14), *and* (2.27)–(2.29).

2.3.5. Formulas (2.13), (2.14), (2.27)–(2.29) give a method for constructing $B_1(\zeta)$ in which it suffices to know the operator-valued function $R(x)$ and J. The point is that $F_1(x)$, $L(x)$, and hence the operator S can be expressed only in terms of $R(x)$ and J. The problem of recovering the Wronskian $W(z)$ from given $R(x)$ and J is of independent interest. Under the assumptions made in Theorem 2.1 formulas (2.13), (2.14), (2.27)–(2.30) provide a solution of this problem. However, a simpler approach is possible here. Indeed, it follows from (2.17), (2.20), and (2.26) that

$$F_2(x) = [S^{-1}F_1(x)]J, \tag{2.35}$$

from which we find $W(z)$ via (2.18) and (2.4). In the case where $J = I$ the problem of recovering $W(z)$ from $R(x)$ plays an essential role in prediction of stationary processes and it was solved by P. Masani and N. Wiener (see [60]).

Chapter 4
Spectral Problems on the Half-line

In this chapter we introduce the notion of a spectral matrix function $\tau(u)$ for a canonical system of differential equations and give a description of the set of all spectral functions of a given system (the direct spectral problem). After that we solve the inverse spectral problem of recovering the corresponding canonical system from $\tau(u)$.

4.1. Basic notions of spectral theory

4.1.1. Let us consider the canonical differential system

$$\frac{dY}{dx} = izJ\mathcal{H}(x)Y, \tag{1.1}$$

where J, $\mathcal{H}(x)$ and $Y(x)$ are $2m \times 2m$, $2m \times 2m$, and $2m \times 1$ matrices, respectively. We assume that

$$J = \begin{bmatrix} 0 & I_m \\ I_m & 0 \end{bmatrix} \quad \text{and} \quad \mathcal{H}(x) \geq 0. \tag{1.2}$$

Let us introduce the space $L^2(\mathcal{H})$ of vector-valued functions $Y(x)$, equipped with the inner product

$$\big(Y(x), Y(x)\big) = \int_0^l Y^*(x)\mathcal{H}(x)Y(x)\,dx. \tag{1.3}$$

If the matrix $\mathcal{H}(x)$ is invertible, then the spectrum of system (1.1) coincides with the spectrum of the operator

$$AY = -i\mathcal{H}^{-1}(x)\,J\,\frac{dY}{dx}, \tag{1.4}$$

acting in the space $L^2(\mathcal{H})$. Here the boundary conditions for problem (1.1) must be taken into account. Later, however, we will drop the requirement that the matrix $\mathcal{H}(x)$ be invertible. To deal with this more general case, we need to develop the spectral theory of the system (1.1) without resorting to the operator A. In important special cases (Sturm-Liouville equation, Dirac-type system) the direct and inverse spectral problems were exhaustively investigated (I. M. Gelfand and B. M. Levitan [25], M. G. Kreĭn [50], V. A. Marchenko [59]). However, for systems more general than the systems mentioned above, there are only results of isolated character and the theory is far from complete (see, e.g., [55], [84]). At the same time, the necessity of investigating new spectral problems is dictated both by mathematical and applied questions.

4.1.2. The vector-valued function $Y(x)$ can be represented in the form

$$Y(x) = \mathrm{col}\,[Y_1(x), Y_2(x)],$$

where $Y_1(x)$, $Y_2(x)$ are vector-valued functions of order $m \times 1$. Let us supplement the system (0.1) by the boundary condition

$$D_2 Y_1(0) + D_1 Y_2(0) = 0, \tag{1.5}$$

where D_1, D_2 are $m \times m$ matrices. We shall assume that

$$D_1 D_2^* + D_2 D_1^* = 0 \tag{1.6}$$

and

$$D_1 D_1^* + D_2 D_2^* = I_m. \tag{1.7}$$

Proposition 1.1. *Let $Y(x)$ be an absolutely continuous vector-valued function which satisfies condition* (1.5). *If relations* (1.6), (1.7) *are fulfilled and $Y(l) = 0$, then*

$$\int_0^l Y^*(x)\, J\, \frac{dY}{dx}\, dx = -\int_0^l \frac{dY^*}{dx}\, J\, Y(x)\, dx. \tag{1.8}$$

Proof. Relation (1.8) is equivalent to

$$Y^*(0) J Y(0) = 0. \tag{1.9}$$

Let us introduce the m-dimensional subspace M defined as the linear span of the columns of the rectangular matrix $\begin{bmatrix} D_2^* \\ D_1^* \end{bmatrix}$. It is easy to see that the boundary condition (1.5) means that the vectors $Y(0)$ are orthogonal to M. Also, condition (1.6) can be interpreted as the orthogonality of JM and M. Therefore,

$$Y(0) \in JM, \qquad JY(0) \in M,$$

which coincides with (1.9). □

4.1.3. Now let us consider the system of integral equations

$$Y(x,z) = Y(0,z) + izJ \int_0^x [dB_1(t)] Y(t,z), \tag{1.10}$$

where $B_1(x)$ is a continuous monotonically increasing $2m \times 2m$ matrix-valued function. If $B_1(x)$ is absolutely continuous, then the system (1.10) is equivalent to the system (1.1), where

$$B_1'(x) = \mathcal{H}(x). \tag{1.11}$$

For the system (1.10) one can introduce the analogue of the space $L^2(\mathcal{H})$, namely, the space $L^2(B_1)$ with the inner product

$$(Y(x), Y(x)) = \int_0^l Y^*(x)[dB_1(x)]Y(x). \tag{1.12}$$

Denote by $\mathcal{W}(x,z)$ the matrix solution of the system of integral equations

$$\mathcal{W}(x,z) = I_{2m} + izJ\int_0^x [dB_1(t)]\mathcal{W}(t,z). \tag{1.13}$$

In the case (1.11), system (1.13) is equivalent to the system

$$\frac{d\mathcal{W}}{dx} = izJ\mathcal{H}(x)\mathcal{W}, \qquad \mathcal{W}(0,z) = I_{2m}. \tag{1.14}$$

Write the block representation

$$\mathcal{W}(x,z) = \left\{\mathcal{W}_{kl}\right\}_{k,l=1}^{2}, \tag{1.15}$$

where the blocks $\mathcal{W}_{kl}(x,z)$ are of order $m \times m$. It follows from (1.6) and (1.7) that

$$UJU^* = J, \tag{1.16}$$

where

$$U = \begin{bmatrix} D_2 & D_1 \\ D_1 & D_2 \end{bmatrix}. \tag{1.17}$$

With system (1.13) and condition (1.5) we associate an operator V by the rule

$$(Vg)(x) = \int_0^l [D_1 \;\; D_2]\mathcal{W}^*(x,u)[dB_1(x)]g(x), \tag{1.18}$$

which maps vector-valued functions $g(x) \in L^2(B_1)$ into vector-valued functions $f(u)$ $(-\infty < u < \infty)$ of order $m \times 1$.

Definition 1.1. A monotonically increasing $m \times m$ matrix-valued function $\tau(u)$ $(-\infty < u < \infty)$ is called *a spectral matrix function* of the system (1.5), (1.13) if the corresponding operator V maps $L^2(B_1)$ isometrically into the space $L^2(\tau)$. (The inner product in $L^2(\tau)$ is defined by

$$(f_1, f_2)_{L^2(\tau)} = \int_{-\infty}^{\infty} f_2^*(u)[d\tau(u)]f_1(u).\ )$$

Definition 1.2. A spectral matrix function $\tau(u)$ is said to be *orthogonal* if the operator V maps $L^2(B_1)$ unitarily *onto* $L^2(\tau)$.

Using the formula

$$(Vg, f)_{L^2(\tau)} = (g, V^* f)_{L^2(B_1)},$$

we deduce that

$$h(x) = V^* f(u) = \int_{-\infty}^{\infty} \mathcal{W}(x,u) \begin{bmatrix} D_1^* \\ D_2^* \end{bmatrix} [d\tau(u)] f(u). \tag{1.19}$$

If $\tau(u)$ is an orthogonal spectral function then

$$V^* f(u) = V^{-1} f(u) = h(x).$$

Let us represent $h(x)$ in the form

$$h(x) = \operatorname{col}[h_1(x), h_2(x)],$$

where $h_1(x)$, $h_2(x)$ are vector-valued functions of order $m \times 1$. It follows from (1.6), (1.19) and the relation $\mathcal{W}(0,u) = I_{2m}$ that the vector $h(x)$ satisfies the boundary condition (1.5).

Theorem 1.1. *The spectral matrix functions of the system* (1.10) *with the boundary condition* (1.5) *coincide with the spectral matrix functions of the system*

$$Y(x,z) = Y(0,z) + izJ \int_0^x [d\widetilde{B}_1(t)] Y(t,z), \tag{1.20}$$

with the boundary condition

$$Y_1(0) = 0, \tag{1.21}$$

where

$$\widetilde{B}_1(x) = UB_1(x)U^*, \qquad U = \begin{bmatrix} D_2 & D_1 \\ D_1 & D_2 \end{bmatrix}. \tag{1.22}$$

Proof. Condition (1.21) is a special case of condition (1.5) if one puts

$$D_1 = 0, \qquad D_2 = I_m. \tag{1.23}$$

The matrix-valued function

$$\widetilde{\mathcal{W}}(x,z) = U^{-1*} \mathcal{W}(x,z) U^* \tag{1.24}$$

and the operator $\widetilde{V}$ acting as

$$(\widetilde{V}g)(u) = \int_0^l [0 \;\; I_m] \widetilde{\mathcal{W}}^*(x,u) [d\widetilde{B}_1(x)] g(x) \tag{1.25}$$

correspond to the boundary value problem (1.20), (1.21). From (1.22) we obtain the equality

$$(g,f)_{B_1} = (U^{*-1}g, U^{*-1}f)_{\widetilde{B}_1} = (\widetilde{V}U^{*-1}g, \widetilde{V}U^{*-1}f)_{\widetilde{\tau}}, \tag{1.26}$$

where $\widetilde{\tau}(u)$ denotes the spectral matrix function of the boundary value problem (1.20), (1.21). From the relation $[D_1 \;\; D_2] = [0 \;\; I_m]U$ and the equalities (1.18), (1.25) it follows that

$$V = \widetilde{V}U^{*-1}. \tag{1.27}$$

Finally, (1.26) yields the relation

$$(g,f)_{B_1} = (Vg, Vf)_{\widetilde{\tau}},$$

which completes the proof of the theorem. □

4.1.4. Let us add to the condition (1.5) at the left endpoint of the interval $(0,l)$ the following condition at the right endpoint l:

$$E_2Y_1(l) + E_1Y_2(l) = 0, \tag{1.28}$$

where E_1 and E_2 are $m \times m$ matrices satisfying

$$E_1E_2^* + E_2E_1^* = 0, \qquad \operatorname{rank}[E_1 \;\; E_2] = m. \tag{1.29}$$

The boundary value problem (1.5), (1.10), (1.29) will be called the *Sturm-Liouville problem*.

Further, let us introduce the matrices

$$D = \begin{bmatrix} D_2 & D_1 \\ 0 & 0 \end{bmatrix}, \qquad E = \begin{bmatrix} 0 & 0 \\ E_2 & E_1 \end{bmatrix}. \tag{1.30}$$

Conditions (1.5) and (1.28) can be recast as

$$DY(0) + EY(l) = 0. \tag{1.31}$$

Definition 1.3. Let g_j be a constant $2m \times 1$ vector, $g_j \neq 0$, and let z_j be a complex number. The vector-valued function

$$Y(x, z_j) = \mathcal{W}(x, z_j)g_j \tag{1.32}$$

is called an *eigenfunction of the system* (1.10), (1.31) if it satisfies the boundary condition

$$DY(0, z_j) + EY(l, z_j) = 0.$$

The corresponding number z_j is called an *eigenvalue* of the boundary value problem (1.10), (1.31).

Using Proposition 1.1 we obtain the following assertion.

Proposition 1.2. *The eigenvalues z_j of the boundary value problem* (1.10), (1.31) *are real. Any two eigenfunctions $Y(x, z_j)$, $Y(x, z_k)$ with $z_j \neq z_k$ are orthogonal in the space $L^2(B_1)$.*

Remark 1.1. The number of linearly independent eigenfunctions corresponding to an eigenvalue z_j is as most $n = 2m$.

Indeed, the number of linearly independent solutions of the equation (1.10) is equal to n for each fixed z.

4.2. Direct and inverse spectral problems

4.2.1. We shall consider the spectral problem for system (1.10) with condition (1.5) when

$$D_1 = 0, \qquad D_2 = I_m. \tag{2.1}$$

Let $\mathfrak{G}_1$ be a vector space with the inner product

$$(g, g)_{\mathfrak{G}_1} = \sum_{k=1}^{m} |g_k|^2, \qquad m < \infty, \tag{2.2}$$

where $g = \mathrm{col}\,[g_1, g_2, \dots, g_m]$.

Then for a symmetric S-node equalities (1.13)–(1.16) of Chapter 2 take the form

$$AS - SA^* = i(\Phi_1\Phi_2^* + \Phi_2\Phi_1^*), \qquad \Pi_1 = [\Phi_1 \ \ \Phi_2]. \tag{2.3}$$

The following theorem is a straightforward consequence of Theorems 1.1 and 1.2 of Chapter 3.

Theorem 2.1. *Suppose that the symmetric S-node (2.2)–(2.3) satisfies the conditions*

I. *The operator S is positive and invertible.*

II. *There exists a continuous monotonically increasing family of orthogonal projectors P_ζ $(0 \le \zeta \le l,\ P_0 = 0,\ P_l = I)$ such that $A^*P_\zeta = P_\zeta A^* P_\zeta$, $0 \le \zeta \le l$.*

III. *The spectrum of the operator A is concentrated at the origin and there is a constant $M > 0$ such that*

$$\|(P_{\zeta+\Delta\zeta} - P_\zeta)A(P_{\zeta+\Delta\zeta} - P_\zeta)\| \le M|\Delta\zeta|, \qquad 0 \le \zeta \le l. \tag{2.4}$$

Then the operator-valued function

$$\mathcal{W}(\zeta, z) = I + izJ\Pi_1^* S_\zeta^{-1}(I - zA_{11})^{-1}P_\zeta\Pi_1, \qquad A_{11} = P_\zeta A P_\zeta, \tag{2.5}$$

admits the representation

$$\mathcal{W}(\zeta, z) = \int_0^{\curvearrowright \zeta} \exp[izJdB_1(t)], \tag{2.6}$$

where

$$B_1(\zeta) = \Pi_1^* S_\zeta^{-1} P_\zeta \Pi_1, \qquad 0 \le \zeta \le l. \tag{2.7}$$

(Notice that $\mathcal{W}(\zeta, z) = W(\zeta, 1/z)$, where $W(\zeta, z)$ is defined by formula (1.10) of Chapter 3.)

It follows from (2.6) that $\mathcal{W}(x, z)$ is a solution of the system of integral equations (1.13).

4.2.2. Denote by $\mathcal{E}_0$ the class of all monotonically increasing matrix-valued functions $\tau(u)$ of order $m \times m$ with the property that the following integral converges weakly:

$$S_\tau = \int_{-\infty}^{\infty} (I - Au)^{-1}\Phi_2[d\tau(u)]\Phi_2^*(I - A^*u)^{-1}. \tag{2.8}$$

An important problem of interpolation theory (see [36]–[38] and [94]) is that of finding a description of those $\tau(u) \in \mathcal{E}_0$ for which the operator S admits the representation

$$S = S_\tau. \tag{2.9}$$

The connection between this interpolation problem and the spectral problem under consideration is clarified by the following theorem.

Theorem 2.2. *Suppose that the conditions of Theorem* 2.1 *are satisfied. If* $\tau(u) \in \mathcal{E}_0$ *and* (2.9) *holds, then* $\tau(u)$ *is a spectral matrix function of the system* (1.10), (2.1).

Proof. Set

$$h_\zeta(x,z) = \begin{cases} \mathcal{W}(x,z)g, & \text{for } x \in [0,\zeta],\ g \in \mathfrak{G} \\ 0, & \text{for } x > \zeta. \end{cases} \tag{2.10}$$

From the equality

$$d\mathcal{W}(x,z) = izJ[dB_1(x)]\mathcal{W}(x,z)$$

and relations (1.13), (1.18) we deduce that

$$Vh_\zeta(x,z) = \frac{i}{z-u}[0\ \ I_m][J - \mathcal{W}^*(\zeta,\bar{u})J\mathcal{W}(\zeta,z)]g. \tag{2.11}$$

In particular, this implies that

$$Vg_\zeta = \frac{i}{u}[0\ \ I_m][\mathcal{W}^*(\zeta,\bar{u}) - I]Jg, \tag{2.12}$$

where

$$g_\zeta = \lim_{z\to 0} h_\zeta(x,z) = \begin{cases} g, & \text{for } x \in [0,\zeta] \\ 0, & \text{for } x > \zeta. \end{cases} \tag{2.13}$$

Thus, by (2.5) and (2.12),

$$Vg_\zeta = \Phi_2^*(I - uA^*)^{-1}S_\zeta^{-1}P_\zeta\Pi_1 g. \tag{2.14}$$

Similarly to g_ζ, let us introduce the function f_η, where $f \in \mathfrak{G}$. Then, by relations (2.8), (2.9) and (2.14),

$$(Vg_\zeta, Vf_\eta)_\tau = (SS_\zeta^{-1}P_\zeta\Pi_1 g, S_\eta^{-1}P_\eta\Pi_1 f)_{\mathfrak{H}} \tag{2.15}$$

where $(Vg_\zeta, Vf_\eta)_\tau$ is the inner product in $L^2(\tau)$. On the other hand, using (2.7) we deduce the equality

$$(g_\zeta, f_\eta)_{B_1} = (SS_\zeta^{-1}P_\zeta\Pi_1 g, S_\zeta^{-1}P_\eta\Pi_1 f)_{\mathfrak{H}}. \tag{2.16}$$

Therefore,

$$(Vg_\zeta, Vf_\eta)_\tau = (g_\zeta, f_\eta)_{B_1}. \tag{2.17}$$

Since g_ζ and f_η form complete systems in $L^2(B_1)$, (2.17) shows that the operator V maps $L^2(B_1)$ isometrically into $L^2(\tau)$. The theorem is proved. □

The following assertion is a partial converse of Theorem 2.2.

Theorem 2.3. *Suppose that the conditions of Theorem* 2.1 *are satisfied and the system of vectors* $S_\zeta^{-1}P_\zeta\Pi_1 g$, *with* $g \in \mathfrak{G} = \mathfrak{G}_1 \oplus \mathfrak{G}_1$, $0 \le \zeta \le l$, *is complete in* $\mathfrak{H}$. *If* $\tau(u)$ *is a spectral matrix function of the system* (1.10), (2.1), *then* $\tau(u) \in \mathcal{E}_0$ *and* (2.9) *holds.*

Proof. As above, equality (2.16) holds. Then, thanks to (2.17), equality (2.15) also holds. The theorem now follows from the completeness of the system $S_\zeta^{-1}P_\zeta\Pi_1 g$ and relation (2.15). □

Corollary 2.1. *Suppose that the conditions of Theorem* 2.1 *are satisfied and that the system of vectors* $S_\zeta^{-1}P_\zeta\Pi_1 g$, *with* $g \in \mathfrak{G}$, $0 \le \zeta \le l$ *is complete in* $\mathfrak{H}$. *Then the set of spectral matrix functions* $\tau(u)$ *of the system* (1.10), (2.1) *coincides with the set of matrix-valued functions of class* $\mathcal{E}_0$ *that give a representation of* S *in the form* (2.9).

4.2.3. To describe the set of all spectral functions of the system (1.10), (2.1) we shall use the description of the set of matrix-valued functions $\tau(u)$ that give the representation (2.9).

Let us denote by $\mathcal{E}$ the class of matrix-valued functions $\tau(u) \in \mathcal{E}_0$ for which the integral

$$I_\tau = \int_{-\infty}^{\infty} \frac{d\tau(u)}{1+u^2} \tag{2.18}$$

converges. We claim that the integral

$$\Phi_{1,\tau} = -i\int_{-\infty}^{\infty} \left[A(I-uA)^{-1} + \frac{u}{1+u^2}I\right]\Phi_2\,d\tau(u) \tag{2.19}$$

converges weakly whenever $\tau(u) \in \mathcal{E}$.

Indeed, from the convergence of the integrals in (2.8) and (2.18) it follows that $\Phi_2^*(I-A^*u)^{-1}f \in L^2(\tau)$ and $g/\sqrt{1+u^2} \in L^2(\tau)$ for all $f \in \mathfrak{H}$ and all $g \in \mathfrak{G}_1$. Observing that

$$A(I-uA)^{-1} + \frac{u}{1+u^2}I = \frac{1}{1+u^2}(A+uI)(I-uA)^{-1},$$

we conclude that the integral (2.19) converges weakly.

Lemma 2.1. *Suppose that the operator* S *satisfies relations* (2.3) *and admits a representation* (2.9), *where* $\tau(u) \in \mathcal{E}$. *If* $\Phi_2 g = 0$ *holds only for* $g = 0$, *then*

$$\Phi_1 = \Phi_{1,\tau} + i\Phi_2\alpha, \qquad \alpha = \alpha^* \in \{\mathfrak{G}_1, \mathfrak{G}_1\}. \tag{2.20}$$

Proof. It follows directly from (2.9) and (2.19) that

$$AS - SA^* = i(\Phi_{1,\tau}\Phi_2^* + \Phi_2\Phi_{1,\tau}^*). \tag{2.21}$$

Comparing relations (2.3) and (2.21), we see that

$$(\Phi_1 - \Phi_{1,\tau})\Phi_2^* + \Phi_2(\Phi_1^* - \Phi_{1,\tau}^*) = 0. \tag{2.22}$$

By the assumption of the lemma, the operator Φ_2 is invertible on the subspace $\mathfrak{H}_1 = \Phi_2\mathfrak{G}_1$. Hence, by (2.22), $\Phi_2^*\mathfrak{H}_1 = \mathfrak{G}_1$ and $(\Phi_1 - \Phi_{1,\tau})\mathfrak{G}_1 \subset \mathfrak{H}_1$. It follows that the operator $\alpha = -i\Phi_2^{-1}(\Phi_1 - \Phi_{1,\tau}) \in \{\mathfrak{G}_1, \mathfrak{G}_1\}$ is well-defined and so relation (2.22) yields the desired equalities (2.20) and $\alpha = \alpha^*$. □

Now consider the matrix-valued function

$$v(z) = \alpha + \int_{-\infty}^{\infty} \left(\frac{1}{u-z} - \frac{u}{1+u^2} \right) d\tau(u), \tag{2.23}$$

where $\tau(u)$ and α are such that relations (2.9) and (2.20) are satisfied. In this manner we associate with S and Φ_1 the set $\mathcal{N}(S, \Phi_1)$ of functions $v(z)$ of form (2.23). It is well known that the matrix-valued functions $v(z)$ are analytic for $\operatorname{Im} z \neq 0$ and satisfy the relations

$$v(z) = v^*(\bar{z}), \qquad \frac{v(z) - v^*(z)}{z - \bar{z}} \geq 0. \tag{2.24}$$

We will describe the set $\mathcal{N}(S, \Phi_1)$ in terms of pairs of matrix-valued functions $R(z)$ and $Q(z)$ meromorphic in the upper complex half-plane. Such a pair $R(z)$, $Q(z)$ is said to be *special* if for any of its points of holomorphy z there exists a vector $e \in \mathfrak{G}_1$, $e \neq 0$, such that

$$[R^*(z)R(z) + Q^*(z)Q(z)]e = 0.$$

We say that a pair $R(z)$, $Q(z)$ has the *J-property* if

$$[R^*(z) \;\; Q^*(z)]J \begin{bmatrix} R(z) \\ Q(z) \end{bmatrix} \geq 0, \qquad \operatorname{Im} z > 0. \tag{2.25}$$

The author joint with T. S. Ivanchenko described the set $\mathcal{N}(S, \Phi_1)$ by means of a linear-fractional transformation acting on pairs $R(z)$, $Q(z)$ (see [36]–[38] and [94]).

Theorem 2.4. *Suppose that a symmetric S-node* (2.3) *satisfies the following conditions:*

I. *The operator S is positive and invertible.*
II. *The spectrum of the operator A is concentrated at the origin.*
III. *Zero is not an eigenvalue of A.*
IV. *The ranges of A and Φ_2 have null intersection.*
V. *$\Phi_2 g = 0$ holds only for $g = 0$.*

Then the functions $v(z)$ in $\mathcal{N}(S, \Phi_1)$ and only they admit the representation

$$v(z) = i[a(z)R(z) + b(z)Q(z)] \cdot [c(z)R(z) + d(z)Q(z)]^{-1}, \qquad \operatorname{Im} z > 0, \quad (2.26)$$

where $R(z)$, $Q(z)$ is a non-special pair with the J-property and where the coefficient matrix of the linear-fractional transformation (2.26) *has the form*

$$\mathfrak{A}(z) = \begin{bmatrix} a(z) & b(z) \\ c(z) & d(z) \end{bmatrix} = E - iz\Pi_1(E - zA^*)^{-1}S^{-1}\Pi_1 J. \quad (2.27)$$

The proof of Theorem 2.4 is contained in the book [94], Ch.1.

4.2.4. Using Theorem 2.4 we can prove the following result.

Theorem 2.5. *Suppose that a symmetric S-node* (2.3) *satisfies the conditions of Theorems* 2.1 *and* 2.4. *Let $v(z)$ be a matrix-valued function of the form* (2.26), *where R and Q are constant matrices and*

$$Q^*R + R^*Q = 0, \qquad R^*R + Q^*Q > 0. \quad (2.28)$$

Then the function $\tau(u)$ in the representation (2.23) *is an orthogonal spectral matrix function of a system* (1.10), (2.1).

Proof. It follows from Theorems 2.2 and 2.4 that $\tau(u)$ is a spectral matrix function. By (2.26)–(2.28), the function $v(z)$ is meromorphic in the complex plane and satisfies (2.24). Hence, $\tau(u)$ is a step matrix-valued function. Denote by z_j ($z_j = \bar{z}_j$) the jump points of $\tau(u)$. Consider the system (1.10) with the boundary condition (1.22), where

$$D = \begin{bmatrix} I_m & 0 \\ 0 & 0 \end{bmatrix}, \qquad E = \begin{bmatrix} 0 & 0 \\ R^* & Q^* \end{bmatrix}. \quad (2.29)$$

It is an immediate consequence of relations (2.27), (2.29) and (1.32) that $Y(x, z_j)$ is an eigenfunction of problem (1.10), (1.22) and (2.29) if and only if $g_j = \operatorname{col}[0, g_{2j}]$ and

$$[R^*c^*(z_j) + Q^*d^*(z_j)]\, g_{2j} = 0, \qquad g_{2j} \neq 0. \quad (2.30)$$

Two eigenfunctions $Y(x, z_j)$ and $Y(x, z_k)$ are orthogonal in $L^2(B_1)$ whenever $z_j \neq z_k$. This fact is equivalent to the relation

$$([d(z_k)c^*(z_j) + c(z_k)d^*(z_j)]g_{2j}, g_{2k}) = 0. \quad (2.31)$$

By (1.32) and (2.11),

$$VY(x, z_j) = \frac{i}{u - z_j}[d(u)c^*(z_j) + c(u)d^*(z_j)]g_{2j}, \quad (2.32)$$

which in conjunction with (2.26) yields

$$[R^*c^*(z_j)+Q^*d^*(z_j)]\gamma_j=0, \tag{2.33}$$

where γ_j denotes the jump of $\tau(u)$ at the point z_j. By relations (2.31) and (2.32),

$$(VY(x,z_j),f(u))_\tau = i(\gamma_j L_j g_{2j}, f(z_j))_{\mathfrak{G}_1}, \tag{2.34}$$

where $f(u)\in L^2(\tau)$ and

$$L_j = d'(z_j)c^*(z_j)+c'(z_j)d^*(z_j). \tag{2.35}$$

Denote by m_j the dimension of the subspace of eigenfunctions corresponding to z_j. By (2.30) and (2.33), $m_j\ge \operatorname{rank}\gamma_j$. Recalling that the operator V is isometric, we conclude that $m_j=\operatorname{rank}\gamma_j$. Hence, the vectors $\gamma_j L_j g_{2j}$ form a complete system in the range of γ_j. Finally, thanks to relation (2.34), the vectors $VY(x,z_j)$ form a complete system in $L^2(\tau)$, that is, $\tau(u)$ is an orthogonal spectral matrix function. The theorem is proved. □

Let us denote by M_j the range of the matrix γ_j. From (2.34) it follows that

$$VY(x,z_j)=i\mathcal{P}_j L_j g_{2j},$$

where $\mathcal{P}_j L_j g_{2j}$ is the orthogonal projection of $L_j g_{2j}$ on M_j. Then the following assertion holds true.

Corollary 2.2. *The operator V maps the eigenfunctions of the boundary value problem* (1.22), (1.29) *into eigenfunctions of the operator Q defined as*

$$Qf=uf(u), \qquad f(u)\in L^2(\tau).$$

Corollary 2.3. *Suppose that the conditions of Theorem* 2.5 *are satisfied. Then the system of eigenfunctions of the system* (1.10), (1.31), (2.29) *is complete in $L^2(B_1)$.*

Corollary 2.4. *Suppose that a symmetric S-node satisfies the conditions of Theorems* 2.1 *and* 2.4. *Then the system of vectors $\{S_\zeta^{-1}P_\zeta\Pi_1 g\}$ ($g\in\mathfrak{G}$, $0\le\zeta\le l$) is complete in $\mathfrak{H}$.*

Proof. Suppose that there exists an $f_0\in\mathfrak{H}$ such that

$$(f_0, S_\zeta^{-1}P_\zeta\Pi_1 g)=0, \qquad 0\le\zeta\le l, \tag{2.36}$$

for all $g\in\mathfrak{G}$. Since the system $\{g_\zeta\}$ is complete in $L^2(B_1)$, it follows that the system $\{Vg_\zeta\}$ is complete in $L^2(\tau)$, where $\tau(u)$ is an orthogonal spectral matrix function generated by a pair R, Q of constant matrices satisfying (2.28) (see Theorem 2.5). Relations (2.8) and (2.9) hold according to Theorem 2.4. Further, by condition II of Theorem 2.1 and (2.14),

$$(f_0,\ S_\zeta^{-1}P_\zeta\Pi_1 g)_{\mathfrak{H}} = (\Phi_2^*(I-A^*u)^{-1}S_\zeta^{-1}P_\zeta f_0, Vg_\zeta)_\tau. \tag{2.37}$$

From the completeness of the system $\{Vg_\zeta\}$ in $L^2(\tau)$ and relations (2.36), (2.37) it follows that

$$(\Phi_2(I-A^*u)^{-1}S_\zeta^{-1}P_\zeta f_0, \Phi_2(I-A^*u)^{-1}S_\zeta^{-1}P_\zeta f_0)_\tau=0. \tag{2.38}$$

Using (2.8) and (2.9) one more time, we deduce from (2.38) that $(f_0,S_\zeta^{-1}f_0)_{\mathfrak{H}}=0$, that is, $f_0=0$. The corollary is proved. □

It follows from Theorems 2.2, 2.3 and Corollary 2.4 that *the set of all spectral matrix functions $\tau(u)$ of the system (1.10), (2.1) coincides with the set of all solutions $\tau(u) \in \mathcal{E}$ of the interpolation problem* (2.9).

With all this information, Theorem 2.4 yields a description of the set of spectral matrix functions $\tau(u)$ of the system (1.10), (2.1).

Theorem 2.6. *Suppose that the symmetric S-node*

$$AS - SA^* = i(\Phi_1\Phi_2^* + \Phi_2\Phi_1^*),$$

where $S, A \in \{\mathfrak{H}, \mathfrak{H}\}, \Phi_1, \Phi_2 \in \{\mathfrak{G}_1, \mathfrak{H}\}$, satisfies the following conditions:

I. *The operator S is positive and invertible.*

II. *There exists a continuous increasing family of orthogonal projectors P_ζ, $(0 \le \zeta \le l,\ P_0 = 0,\ P_l = E)$ such that $A^*P_\zeta = P_\zeta A^* P_\zeta$, $0 \le \zeta \le l$.*

III. *The spectrum of the operator A is concentrated at the origin and there exist a constant $M > 0$ such that*

$$\|(P_{\zeta+\Delta\zeta} - P_\zeta)A(P_{\zeta+\Delta\zeta} - P_\zeta)\| \le M|\Delta\zeta|, \qquad 0 \le \zeta \le l.$$

IV. *Zero is not an eigenvalue of A.*

V. *The ranges of A and Φ_2 have null intersection.*

VI. *$\Phi_2 g = 0$ implies $g = 0$.*

Then the following assertions hold:

(i) *If $v(z)$ admits a representation (2.26), then the function $\tau(u)$ figuring in the formula*

$$v(z) = \alpha + \int_{-\infty}^{\infty} \left(\frac{1}{u-z} - \frac{u}{1+u^2}\right) d\tau(u) \tag{2.39}$$

is a spectral matrix function of the system

$$Y(x,z) = Y(0,z) + izJ\int_0^x [dB_1(t)]Y(t,z) \tag{2.40}$$

with the boundary condition

$$Y_1(0) = 0 \qquad \big(Y(x,z) = \operatorname{col}[Y_1(x,z), Y_2(x,z)]\big). \tag{2.41}$$

(ii) *If $\tau(u)$ is a spectral matrix function of the system (2.40) with the boundary condition (2.41) and if*

$$I_\tau = \int_{-\infty}^{\infty} \frac{d\tau(u)}{1+u^2} < \infty, \tag{2.42}$$

then there exists a matrix $\alpha = \alpha^* \in \{\mathfrak{G}_1, \mathfrak{G}_1\}$ such that the matrix $v(z)$ defined by (2.39) admits the representation (2.26).

To make this statement complete, recall that the matrix $B_1(t)$ figuring in formula (2.40) is defined by the relation

$$B_1(\zeta) = \Pi^* S_\zeta^{-1} P_\zeta \Pi, \tag{2.43}$$

where $\Pi = [\Phi_1, \Phi_2]$, $S_\zeta = P_\zeta S P_\zeta$.

4.2.5. Formulas (2.7), (2.8), and (2.19) allow us to recover the system (2.40) if we are given $\tau(u)$ and α, that is, to solve the inverse spectral problem. We obtain the following result.

Theorem 2.7. *Suppose that the operators A and Φ_2 satisfy conditions* II–VI *of Theorem* 2.6. *If a matrix function $\tau(u)$ in the class $\mathcal{E}$ is such that the operator S_τ defined by* (2.8) *is bounded with bounded inverse S_τ^{-1}, then $\tau(u)$ is a spectral matrix function of a system* (2.40), (2.41), *which is defined by* (2.7), (2.8), (2.19) *and the equalities*

$$\Pi_1 = [\Phi_1 \ \ \Phi_2], \qquad S = S_\tau, \qquad \Phi_1 = \Phi_{1,\tau} + i\Phi_2\alpha. \tag{2.44}$$

Remark 2.1. Fix a matrix-valued function $\tau(u)$ satisfying the conditions of Theorem 2.7. Then formulas (2.7), (2.44) associate different α with different $B_1(\zeta,\alpha)$ and $\mathcal{W}(x,z,\alpha)$. Specifically, these formulas yield the following dependence of $B_1(\zeta,\alpha)$ on α:

$$B_1(\zeta,\alpha) = T(\alpha)B_1(\zeta,0)T^*(\alpha), \tag{2.45}$$

where $T(\alpha)$ has the block decomposition

$$T(\alpha) = \begin{bmatrix} I_m & -i\alpha \\ 0 & I_m \end{bmatrix}. \tag{2.46}$$

Since $T^*(\alpha)JT(\alpha) = J$, we deduce from (1.13) that

$$\mathcal{W}(x,z,\alpha) = T^{*-1}(\alpha)\mathcal{W}(x,z,0)T^*(\alpha). \tag{2.47}$$

4.3. Livšic-Brodskiĭ nodes and the spectral theory of canonical systems

4.3.1. If in relation (2.3) we take for S the operator

$$S = I, \tag{3.1}$$

then the operators A, Φ_1, Φ_2 form a *Livšic-Brodskiĭ node* (see [14], Ch. 1). Under condition (3.1) identity (2.3) becomes

$$A - A^* = i(\Phi_1\Phi_2^* + \Phi_2\Phi_1^*). \tag{3.2}$$

In this case Theorem 2.1 admits the following reformulation.

Theorem 3.1. *Assume that* (3.2) *holds and the following conditions are satisfied:*

a) *There exists an increasing continuous family of orthogonal projectors* P_ζ $(0 \leq \zeta \leq l,\ P_0 = 0, P_l = I)$, *such that*

$$A^*P_\zeta = P_\zeta A^* P_\zeta, \qquad 0 \leq \zeta \leq l.$$

b) *The spectrum* A *reduces to* $\{0\}$ *and there exists a constant* $M > 0$ *such that*

$$\|(P_{\zeta+\Delta\zeta} - P_\zeta)A(P_{\zeta+\Delta\zeta} - P_\zeta)\| \leq M|\Delta\zeta|.$$

Then the matrix-valued function

$$W(\zeta, z) = I_n + izJ\Pi^*(I - zA_\zeta)^{-1}P_\zeta\Pi \tag{3.3}$$

admits the multiplicative integral representation

$$W(\zeta, z) = \overset{\curvearrowleft}{\int_0^\zeta} \exp[izJdB_1(t)], \tag{3.4}$$

where

$$\Pi = [\Phi_1 \ \ \Phi_2], \qquad A_\zeta = P_\zeta A P_\zeta, \qquad B_1(\zeta) = \Pi^* P_\zeta \Pi. \tag{3.5}$$

Similarly, in the case (3.1) Theorem 2.4 admits the following reformulation.

Theorem 3.2. *Let in addition to the assumptions of Theorem* 3.1 *the following conditions be satisfied:*

c) *Zero is not an eigenvalue of* A.

d) *The ranges of* A *and* Φ_2 *have null intersection;*

e) $\Phi_2 g = 0$ *implies* $g = 0$.

Then the functions $v(z) \in \mathcal{N}(S, \Phi_1)$ *and only they admit the representation* (2.26), (2.27).

4.3.2. In the remaining part of this section we will assume the matrix-valued function $B_1(t)$ is absolutely continuous and satisfies $\operatorname{Tr} B_1'(t) \equiv 1$. This normalization can be always achieved by the choosing an appropriate parametrization of the family of orthogonal projectors P_ζ. Then thanks to (3.4) one has the equality

$$\frac{dW(x,z)}{dx} = izJ\mathcal{H}(x)W(x,z), \tag{3.6}$$

where

$$\mathcal{H}(x) = B_1'(x), \qquad \operatorname{Tr} \mathcal{H}(x) \equiv 1.$$

Thus, starting with the operator identity (3.2) we have produced a system (3.6) and have described the set of all spectral matrix functions of this system.

4.3.3. Now let us take as starting point a system of the form (3.6). With this system we associate the model operator A_{M}, which acts according to the rule

$$(A_{\mathrm{M}}f)(x) = i\beta(x)J\int_0^x \beta^*(t)f(t)\,dt, \tag{3.8}$$

where $f \in L_r^2(0,l)$, $\beta(t)$ is an $r \times n$ matrix-valued function and

$$\beta^*(x)\beta(x) = \mathcal{H}(x), \qquad r = \sup_{0\le x\le l} \operatorname{rank}\mathcal{H}(x). \tag{3.9}$$

Relation (3.8) implies the identity

$$(A_{\mathrm{M}} - A_{\mathrm{M}}^*)f = i\beta(x)J\int_0^l \beta^*(t)f(t)\,dt. \tag{3.10}$$

Let us write the matrix $\beta(x)$ in the block form

$$\beta(x) = [\beta_1(x)\ \ \beta_2(x)],$$

where $\beta_1(x)$, $\beta_2(x)$ are matrices of order $r \times m$ ($n = 2m$).

Now set

$$\Phi_k g = \beta_k(x)g, \qquad g \in \mathfrak{G}_1,\ \ k = 1,2. \tag{3.11}$$

Then identity (3.10) becomes

$$A_{\mathrm{M}} - A_{\mathrm{M}}^* = i(\Phi_1\Phi_2^* + \Phi_2\Phi_1^*). \tag{3.12}$$

Further, let us introduce a family of orthogonal projectors in the space $L_r^2(0,l)$ by setting

$$P_\zeta f = \begin{cases} f(x), & \text{for } 0 \le x \le \zeta \le l \\ 0, & \text{for } \zeta \le x > l. \end{cases} \tag{3.13}$$

The operators A_{M}, Φ_1, Φ_2, and P_ζ satisfy all the assumptions of Theorem 3.1. By (3.5), (3.9), and (3.11), we have

$$B_1'(x) = \beta^*(x)\beta(x) = \mathcal{H}(x),$$

that is, the system (3.6) under consideration corresponds to the operator identity (3.12).

Theorem 3.2 admits the following consequence.

Theorem 3.3. *Suppose the operator A_{M} generated by $\mathcal{H}(x)$ according to the recipes* (3.8) *and* (3.9) *satisfies the following conditions:*

a) *Zero is not an eigenvalue of A_{M};*

b) *The vector $\beta_2(x)g$ does not belong to the range of A_{M} if $g \neq 0$;*

c) *$\beta_2(x)g = 0$ implies $g = 0$.*

Then the set of matrix-valued functions $v(z)$ of the corresponding canonical system (3.6) *is described by means of the formula*

$$v(z) = i[a(z)R(z) + b(z)Q(z)][c(z)R(z) + d(z)Q(z)]^{-1}, \qquad \mathrm{Im} z > 0, \tag{3.14}$$

where $R(z)$, $Q(z)$ is a non-special pair with the J-property and the matrix of coefficients of the linear-fractional transformation (3.14) has the form

$$\mathfrak{A}(z) = \begin{bmatrix} a(z) & b(z) \\ c(z) & d(z) \end{bmatrix} = I - iz\Pi_1^*(I - zA_M^*)^{-1}\Pi_1 J, \qquad \Pi_1 = [\Phi_1 \ \ \Phi_2] \ . \tag{3.15}$$

In this section we have passed to the case $S = I$ and as a result a number of statements have become simpler (compare Theorems 2.6 and 3.3). However this simplification takes place only for the direct spectral problem (description of the set of spectral matrix functions $\tau(u)$). In the procedure for solving the inverse spectral problem (finding $\mathcal{H}(x)$ when $\tau(u)$ is given) it is essential that the operator A is fixed. Generally speaking, in this case the operator S is different from the identity operator. An important step in solving the inverse problem is constructing the operator S by means of the formula

$$S = \int_{-\infty}^{\infty} (I - Au)^{-1}\Phi_2[d\tau(u)]\Phi_2^*(I - A^*u)^{-1}.$$

Proposition 3.1. *Condition* a) *of Theorem* 3.3 *is satisfied whenever the function $\|\beta'(\cdot)\|$ belongs to $L^2(0,l)$ and*

$$\det[\beta(x)J\beta^*(x)] \neq 0, \qquad 0 \leq x \leq l. \tag{3.16}$$

Proof. Assume the equality

$$\beta(x)\int_0^x J\beta^*(t)f(t)\,dt = 0 \tag{3.17}$$

holds. Differentiating both sides of (3.17), we get

$$\beta(x)J\beta^*(x)f(x) + \int_0^x \beta'(x)J\beta^*(t)f(t)\,dt = 0,$$

which thanks to (3.16) implies that $f(x) = 0$. The proposition is proved. □

Remark 3.1. Conditions b) and c) of Theorem 3.3 are satisfied whenever $\beta_2(x)$ is continuous at the point $x = 0$ and

$$\operatorname{rank} \beta_2(0) = m. \tag{3.18}$$

Indeed, from relation (3.18) it follows that

$$\beta_2(0)g \neq 0 \qquad \text{when} \quad g \neq 0. \tag{3.19}$$

Recalling formula (3.8), we obtain

$$(A_{\mathrm{M}} f)\big|_{x=0} = 0, \tag{3.20}$$

which in conjunction with (3.19) shows that conditions b) and c) of Theorem 3.3 are indeed satisfied.

Chapter 5
Spectral Problems on the Line

For canonical systems of differential equations spectral problems on the line $(-\infty, +\infty)$ can be reduced to problems on the half-line $(0, \infty)$. As in the classical case of the Sturm-Liouville equation, such a reduction is performed by doubling the dimension of the system.

5.1. Spectral data of a canonical system

5.1.1. Let us consider two Hilbert spaces $\mathfrak{G}$ and $\mathfrak{H}$, where

$$\mathfrak{G} = \mathfrak{G}_1 \oplus \mathfrak{G}_1, \qquad \dim \mathfrak{G}_1 = m < \infty. \tag{1.1}$$

We will assume that the matrix J has the form

$$J = \begin{bmatrix} 0 & I_m \\ I_m & 0 \end{bmatrix} \tag{1.2}$$

and that $B(x)$ is a continuous monotonically increasing $2m \times 2m$ matrix-valued function on the segment $[-l, l]$. Consider the system of integral equations

$$\mathcal{W}(x, z) = I_{2m} + izJ \int_0^x [dB(t)]\mathcal{W}(t, z). \tag{1.3}$$

If $B(x)$ is absolutely continuous, then (1.3) is equivalent to the canonical system

$$\frac{d\mathcal{W}(x, z)}{dx} = izJ\mathcal{H}(x)\mathcal{W}(x, z), \tag{1.4}$$

where

$$\mathcal{W}(0, z) = I_{2m}, \quad \mathcal{H}(x) = B'(x).$$

Denote by $L^2(B)$ the space of column-vectors $g(x)$, equipped with the inner product

$$(g, h) = \int_{-l}^{l} h^*(x)[dB(x)]g(x). \tag{1.5}$$

A monotonically increasing $2m \times 2m$ matrix-valued function $\tau(u)$ $(-\infty < u < \infty)$ is called a *spectral matrix function* of the system (1.3) if the operator V acting by the rule

$$(Vg)(u) = \frac{1}{\sqrt{2}} \int_{-l}^{l} \mathcal{W}^*(x, u)[dB(x)]g(x) = F(u). \tag{1.6}$$

maps $L^2(B)$ isometrically into $L^2(\tau)$. Here $F(u)$ is a $2m$-dimensional column vector, and the space $L^2(\tau)$ is defined by means of the inner product

$$(F_1, F_2) = \int_{-\infty}^{\infty} F_2^*(u)[d\tau(u)]F_1(u). \tag{1.7}$$

A spectral matrix function $\tau(u)$ is said to be *orthogonal* if the operator V maps $L^2(B)$ isometrically *onto* $L^2(\tau)$. Using the identity

$$(Vg, F)_{L^2(\tau)} = (g, V^*F)_{L^2(B)},$$

we deduce that

$$(V^*F)(x) = \frac{1}{\sqrt{2}} \int_{-\infty}^{\infty} \mathcal{W}(x,u)[d\tau(u)]F(u). \tag{1.8}$$

If $\tau(u)$ is an orthogonal spectral matrix function, then the equality

$$V^*F = V^{-1}F \tag{1.9}$$

holds for all $F \in L^2(\tau)$.

5.1.2. In order to reduce the spectral problem on the line to the problem on the half-line investigated earlier (Chapter 4), we introduce the following notations:

$$\mathcal{W}_+(x,z) = \mathcal{W}(x,z), \quad \mathcal{W}_-(x,z) = \mathcal{W}(-x,z), \quad x \geq 0, \tag{1.10}$$

$$B_+(x) = B(x), \quad B_-(x) = -B(-x), \quad x \geq 0. \tag{1.11}$$

Set

$$B_0(x) = T^{-1} \begin{bmatrix} B_+(x) & 0 \\ 0 & B_-(x) \end{bmatrix} T,$$

$$\mathcal{W}_0(x,z) = T^{-1} \begin{bmatrix} \mathcal{W}_+(x,z) & 0 \\ 0 & \mathcal{W}_-(x,z) \end{bmatrix} T, \tag{1.12}$$

$$T = \frac{1}{\sqrt{2}} \begin{bmatrix} J & I_{2m} \\ -J & I_{2m} \end{bmatrix}, \qquad J_0 = \begin{bmatrix} 0 & I_{2m} \\ I_{2m} & 0 \end{bmatrix}. \tag{1.13}$$

We can now rewrite (1.3) in the form

$$\mathcal{W}_0(x,z) = I_{4m} + izJ_0 \int_0^x [dB_0(t)]\mathcal{W}_0(t,z), \qquad 0 \leq x \leq l. \tag{1.14}$$

(If $B_0(x)$ is absolutely continuous then the system (1.14) is equivalent to the canonical system

$$\frac{d\mathcal{W}_0(x,z)}{dx} = izJ_0\mathcal{H}_0(x)\mathcal{W}_0(x,z),$$

where $\mathcal{W}(0,z) = I_{4m}$, $\mathcal{H}_0(x) = B_0'(x)$.)

Next, let us introduce the operator V_0 acting by the rule

$$(V_0 g)(u) = \int_0^l [0 \;\; I_{2m}]\mathcal{W}_0^*(x,u)[dB_0(x)]g(x) = f(u); \tag{1.15}$$

here the vector $f(u)$ has dimension $2m$.

A monotonically increasing $2m \times 2m$ matrix-valued function $\tau(u)$ $(-\infty < u < \infty)$ is a spectral matrix function for the system (1.14) if the operator V_0 maps $L^2(B_0)$ isometrically into $L^2(\tau)$ (see Chapter 4.1).

Thus, starting with the system (1.3) on the segment $[-l, l]$ we arrived at a system (1.14) of doubled dimension on the segment $[0, l]$. The connection between these systems is clarified further by the following theorem.

Theorem 1.1. *The families of spectral matrix functions* $\tau(u)$ *of the system* (1.3) *and of the corresponding system* (1.14) *coincide.*

Proof. Consider the following inner product:

$$\begin{aligned}(g,h) = \int_{-l}^{l} h^*(x)[dB(x)]g(x) = \\ = \int_0^l h_1^*(x)[dB_+(x)]g_1(x) + \int_0^l h_2^*(x)[dB_-(x)]g_2(x),\end{aligned} \tag{1.16}$$

where

$$h_1(x) = h(x), \quad h_2(x) = h(-x), \quad g_1(x) = g(x), \quad g_2(x) = g(-x), \quad 0 \le x \le l.$$

Next, set

$$\tilde{h} = \begin{bmatrix} h_1 \\ h_2 \end{bmatrix}, \qquad \tilde{g} = \begin{bmatrix} g_1 \\ g_2 \end{bmatrix}$$

and rewrite (1.16) in the form

$$(g,h) = \int_0^l \tilde{h}^*(x)T[dB_0(x)]T^{-1}\tilde{g}(x) = \int_0^l [T^{-1}\tilde{h}(x)]^*[dB_0(x)][T^{-1}\tilde{g}(x)]. \tag{1.17}$$

Now let us find the relationship between V and V_0. By (1.6) and (1.10)–(1.12), one has the equality

$$Vg = \frac{1}{\sqrt{2}}\int_0^l \{\mathcal{W}_+^*(x,u)[dB_+(x)]g_1(x) + \mathcal{W}_-^*(x,u)[dB_-(x)]g_2(x)\},$$

which in turn can be recast as

$$Vg = \frac{1}{\sqrt{2}}\int_0^l [\mathcal{W}_+^*(x,u) \;\; \mathcal{W}_-^*(x,u)]T[dB_0(x)][T^{-1}\tilde{g}(x)]. \tag{1.18}$$

Thanks to relations (1.12), (1.13), (1.15), and (1.18), we have

$$Vg = V_0(T^{-1}\tilde{g}). \tag{1.19}$$

The assertion of the theorem follows from (1.17) and (1.19). □

5.1.3. We note that, by (1.12) and (1.13), the matrix $\mathcal{W}_0^*(l,\bar{z})$ admits the block representation

$$\mathcal{W}_0^*(l,\bar{z}) = \frac{1}{2}\begin{bmatrix} Ja_0(l,z)J & Jb_0(l,z) \\ b_0(l,z)J & a_0(l,z) \end{bmatrix}, \tag{1.20}$$

where we use the following notations:

$$a_0(l,z) = \mathcal{W}_+^*(l,\bar{z}) + \mathcal{W}_-^*(l,\bar{z}), \quad b_0(l,z) = \mathcal{W}_+^*(l,\bar{z}) - \mathcal{W}_-^*(l,\bar{z}). \tag{1.21}$$

Now consider the class of matrix-valued functions $v(z)$ of the form

$$v(z) = i[Ja_0(l,z)JR(z) + Jb_0(l,z)Q(z)][b_0(l,z)JR(z) + a_0(l,z)Q(z)]^{-1}, \tag{1.22}$$

where $R(z)$, $Q(z)$ are meromorphic matrix-valued functions such that

$$\det[R^*(z)R(z) + Q^*(z)Q(z)] \not\equiv 0, \tag{1.23}$$

$$R^*(z)Q(z) + Q^*(z)R(z) \geq 0, \qquad \operatorname{Im} z > 0. \tag{1.24}$$

Notice that the matrix of coefficients of the linear-fractional transformation (1.22) has the form

$$\mathfrak{A}_0(l,z) = \mathcal{W}_0^*(l,\bar{z}) = \frac{1}{2}\begin{bmatrix} Ja_0(l,z)J & Jb_0(l,z) \\ b_0(l,z)J & a_0(l,z) \end{bmatrix}.$$

The functions of the form (1.22) belong to the *Nevanlinna class*, i.e., they satisfy

$$\frac{v(z) - v^*(z)}{z - \bar{z}} \geq 0, \qquad \operatorname{Im} z > 0. \tag{1.25}$$

As members of this class, the matrix-valued functions $v(z)$ admits the representation

$$v(z) = \gamma z + \alpha + \int_{-\infty}^{\infty} \left(\frac{1}{u-z} - \frac{u}{1+u^2} \right) d\tau(u), \tag{1.26}$$

where $\gamma \geq 0$, $\alpha = \alpha^*$, and $\tau(u)$ is a monotonically increasing matrix-valued function. When $\gamma = 0$ equality (1.26) becomes

$$v(z) = \alpha + \int_{-\infty}^{\infty} \left(\frac{1}{u-z} - \frac{u}{1+u^2} \right) d\tau(u). \tag{1.27}$$

Under certain additional conditions, the matrix-valued functions $\tau(u)$ figuring in the representation (1.26) are spectral matrix functions for systems of the form (1.14) (see Chapter 4, Section 3). Hence, they are also spectral matrix functions for systems of the form (1.3.).

Let us formulate a corresponding result for the case when $B_0(x)$ is an absolutely continuous matrix-valued function. In this case the system (1.14) can be rewritten in the form

$$\frac{d\mathcal{W}_0(x,z)}{dx} = izJ_0\mathcal{H}_0(x)\mathcal{W}(x,z), \quad \mathcal{W}_0(0,z) = I_{4m}, \tag{1.28}$$

where $\mathcal{H}_0(x) = B_0'(x) \geq 0$. The matrix $\mathcal{H}_0(x)$ can be represented in the form

$$\mathcal{H}_0(x) = \beta_0^*(x)\beta_0(x), \tag{1.29}$$

where $\beta_0(x)$ is an $r \times 4m$ matrix-valued function, with $r = \sup_{0\leq x\leq l} \operatorname{rank} \mathcal{H}_0(x)$. Let us write $\beta_0(x)$ in the block form

$$\beta_0(x) = [\beta_{01}(x) \ \ \beta_{02}(x)]$$

where $\beta_{0k}(x)$ are $r \times 2m$-matrices.

Now let us associate with the system (1.28) the operator

$$A_0 f = \beta_0(x)J\int_0^x \beta_0^*(t)f(t)\,dt, \tag{1.30}$$

where $f(x) \in L_r^2(0,l)$.

Theorem 4.33 admits the following immediate consequence.

Theorem 1.2. *Let the following conditions be satisfied:*

a) *Zero is not an eigenvalue of the operator A_0.*

b) *The vector-valued function $\beta_{02}(x)g$ do not belong to the range of the operator A_0 if $g \neq 0$.*

c) *If $g \not\to 0$, then $\beta_{02}(x)g \not\to 0$.*

Then the set of Weyl-Titchmarsh matrix functions $v(z)$ of the corresponding system (1.3) *is described by means of formula* (1.22).

Proposition 1.1. *Assume that*

$$\operatorname{rank} \beta_0(x) = 4m \tag{1.31}$$

for almost every $x \in [0,l]$. Then condition a) *of Theorem* 1.2 *is satisfied.*

Proof. Indeed, if $A_0 f_0 = 0$ then, by (1.31),

$$\int_0^x \beta_0^*(t)f_0(t)\,dt = 0,$$

i.e., $\beta_0(t)f_0(t) = 0$, whence $f_0(t) = 0$. The proposition is proved. □

Proposition 1.2. *If the matrix $\beta_{02}(x)$ is continuous at $x = 0$ and*

$$\operatorname{rank} \beta_{02}(0) = 2m \tag{1.32}$$

then conditions b) *and* c) *of Theorem* 1.2 *are satisfied.*

Proof. It follows from condition (1.32) that

$$\beta_{02}(0)g \neq 0 \qquad \text{if } g \neq 0. \tag{1.33}$$

The fact that conditions b) and c) hold is an immediate consequence of Theorem 1.2. □

Note that the matrix-valued functions $\beta_0(x)$ can be chosen in the following way:

$$\beta_0(x) = \begin{bmatrix} \sqrt{B'_+(x)} & 0 \\ 0 & \sqrt{B'_-(x)} \end{bmatrix} T.$$

In this case we have

$$\beta_{01}(x) = \frac{1}{\sqrt{2}} \begin{bmatrix} \sqrt{B'_+(x)}J \\ -\sqrt{B'_-(x)}J \end{bmatrix}, \qquad \beta_{02}(x) = \frac{1}{\sqrt{2}} \begin{bmatrix} \sqrt{B'_+(x)} \\ \sqrt{B'_-(x)} \end{bmatrix}.$$

5.2. Spectral problems and S-nodes

In order to obtain an effective procedure for solving the inverse problem on the whole line $(-\infty, \infty)$ we will resort to the theory of S-nodes. We will denote by $\{\mathfrak{H}_1, \mathfrak{H}_2\}$ the set of all bounded operators acting from $\mathfrak{H}_1$ to $\mathfrak{H}_2$.

Let the Hilbert spaces $\mathfrak{H}_\pm$ and the operators $A_\pm$, $S_\pm \in \{\mathfrak{H}_\pm, \mathfrak{H}_\pm\}$, $\Phi_k^\pm \in \{\mathfrak{G}_1, \mathfrak{H}_\pm\}$, $k = 1, 2$, with $\dim \mathfrak{G}_1 = m < \infty$, form S-nodes, i.e., the following relations hold:

$$A_\pm S_\pm - S_\pm A_\pm^* = i[\Phi_1^\pm(\Phi_2^\pm)^* + \Phi_2^\pm(\Phi_1^\pm)^*]. \tag{2.1}$$

Further, assume that the following conditions are satisfied:

1) *The operators $S_\pm$ are bounded, positive and invertible.*
2) *In $\mathfrak{H}_+$ and $\mathfrak{H}_-$ there are continuous monotonically increasing families of orthogonal projectors P_ξ^+ and P_ξ^-, respectively, such that*

$$A_\pm P_\xi^\pm = P_\xi^\pm A_\pm^* P_\xi^\pm, \qquad P_0 = 0, \quad P_l^\pm = I_\pm,$$

where $I_\pm$ is the identity operator in $\mathfrak{H}_\pm$.

3) *The spectra of the operators A_+ and A_- are concentrated in zero and there exists a constant $M > 0$ such that*

$$\|(P_{\xi+\Delta\xi}^\pm - P_\xi^\pm)A_\pm(P_{\xi+\Delta\xi}^\pm - P_\xi^\pm)\| \leq M|\Delta\xi|, \qquad 0 \leq \xi \leq l.$$

Now set

$$S_\xi^\pm = P_\xi^\pm S_\pm P_\xi^\pm, \qquad \Pi_\pm = [\Phi_1^\pm \ \Phi_2^\pm], \tag{2.2}$$

$$B_\pm(\xi) = \Pi_\pm^*(S_\xi^\pm)^{-1}P_\xi\Pi_\pm. \tag{2.3}$$

Then the matrix-valued functions $B_\pm(\xi)$ are continuous and monotonically increasing (see Chapter 3, formula (1.15)).

Let us denote by $\mathcal{W}_\pm(x,z)$ the solutions of the systems

$$\mathcal{W}_\pm(x,z) = I_{2m} \pm izJ\int_0^l [dB_\pm(t)]\mathcal{W}_\pm(t,z). \tag{2.4}$$

Using the notations (1.12) and (1.13), we rewrite (2.4) in the form (1.14). Next, we construct S-nodes corresponding to the system (1.14). To this end we introduce the operators

$$A = \begin{bmatrix} A_+ & 0 \\ 0 & A_- \end{bmatrix}, \qquad S = \begin{bmatrix} S_+ & 0 \\ 0 & S_- \end{bmatrix}, \qquad P_\xi = \begin{bmatrix} P_\xi^+ & 0 \\ 0 & P_\xi^- \end{bmatrix} \tag{2.5}$$

acting in the space $\mathfrak{H} = \mathfrak{H}_+ \oplus \mathfrak{H}_-$. We will also need the operator

$$\Pi_0 = \frac{1}{\sqrt{2}}\begin{bmatrix} \Pi_+ J & \Pi_+ \\ -\Pi_- J & \Pi_- \end{bmatrix}. \tag{2.6}$$

Straightforward calculations show that

$$AS - SA^* = i\Pi_0 J_0 \Pi_0^*, \qquad A^*P_\xi = P_\xi A^* P_\xi. \tag{2.7}$$

Using relations (1.11), (1.12), (2.2), and (2.6), we derive the equality

$$B_0(\xi) = \Pi_0^* S_\xi^{-1} P_\xi \Pi_0. \tag{2.8}$$

It follows from (2.6) that the operators

$$\Phi_1^0 = \frac{1}{\sqrt{2}}\begin{bmatrix} \Phi_2^+ & \Phi_1^+ \\ -\Phi_2^- & -\Phi_1^- \end{bmatrix}, \qquad \Phi_2^0 = \frac{1}{\sqrt{2}}\begin{bmatrix} \Phi_1^+ & \Phi_2^+ \\ \Phi_1^- & \Phi_2^- \end{bmatrix}, \qquad \Pi_0 = [\Phi_1^0 \ \Phi_2^0] \tag{2.9}$$

correspond to the spectral problem (1.14) with the matrix $B_0(x)$ given by (2.8).

When we solved the inverse problem on the half-line, we assumed that the operators A and Φ_2 were fixed. The choice of these operators characterizes the class of the systems for which the inverse problem is being solved. The procedure described Chapter 4 cannot be transferred directly to the case of the system (2.4), because Φ_2^0 depends not only on the fixed $\Phi_2^\pm$, but also on $\Phi_1^\pm$.

In order to fit the present situation into the general scheme we introduce the operators

$$\widetilde{\Phi}_1 = \frac{1}{\sqrt{2}}\begin{bmatrix} \Phi_1^+ & \Phi_1^+ \\ \Phi_1^- & -\Phi_1^- \end{bmatrix}, \qquad \widetilde{\Phi}_2 = \frac{1}{\sqrt{2}}\begin{bmatrix} \Phi_2^+ & \Phi_2^+ \\ -\Phi_2^- & \Phi_2^- \end{bmatrix}, \tag{2.10}$$

$$u_1 = \begin{bmatrix} I_m & 0 \\ 0 & 0 \end{bmatrix}, \qquad u_2 = \begin{bmatrix} 0 & 0 \\ 0 & I_m \end{bmatrix}, \qquad U = \begin{bmatrix} u_2 & u_1 \\ u_1 & u_2 \end{bmatrix}. \tag{2.11}$$

It is readily verified that

$$\Pi_0 = \widetilde{\Pi}U, \qquad UJ_0U^* = J_0, \tag{2.12}$$

where

$$\widetilde{\Pi} = [\widetilde{\Phi}_1 \ \widetilde{\Phi}_2]. \tag{2.13}$$

It follows from (2.7) and (2.12) that the operator identity

$$AS - SA^* = i\widetilde{\Pi}J_0\widetilde{\Pi}^* \tag{2.14}$$

holds for the corresponding operator-valued function

$$\widetilde{B}(\xi) = \widetilde{\Pi}^* S_\xi^{-1} P_\xi \widetilde{\Pi} \tag{2.15}$$

and the system

$$\widetilde{\mathcal{W}}(x,z) = I_{4m} + izJ_0 \int_0^x [d\widetilde{B}(t)]\widetilde{\mathcal{W}}(t,z). \tag{2.16}$$

Equalities (2.8), (2.12) and (2.15) yield

$$B_0(\xi) = U^*\widetilde{B}(\xi)U. \tag{2.17}$$

Comparing the systems (1.14) and (2.16) with the help of relation (2.17), we conclude that

$$\mathcal{W}_0(x,z) = U^{-1}\widetilde{\mathcal{W}}(x,z)U. \tag{2.18}$$

5.3. The inverse spectral problem

5.3.1. Let us associate with the operator identity (2.7) the inequality

$$L(z) = \begin{bmatrix} S & B(z) \\ B^*(z) & C(z) \end{bmatrix} \geq 0, \tag{3.1}$$

where

$$B(z) = (I - Az)^{-1}[\Phi_1^0 - i\Phi_2^0 v(z)], \tag{3.2}$$

and

$$C(z) = \frac{v(z) - v^*(z)}{z - \bar{z}}. \tag{3.3}$$

Inequality (3.1) is an abstract analogue of the Potapov matrix inequality. Let us also consider the inequality

$$\widetilde{L}(z) = \begin{bmatrix} S & \widetilde{B}(z) \\ \widetilde{B}^*(z) & \widetilde{C}(z) \end{bmatrix} \geq 0, \tag{3.4}$$

where

$$\widetilde{B}(z) = (I - Az)^{-1}[\widetilde{\Phi}_1 - i\widetilde{\Phi}_2\tilde{v}(z)], \tag{3.5}$$

and

$$\widetilde{C}(z) = \frac{\tilde{v}(z) - \tilde{v}^*(z)}{z - \bar{z}}. \tag{3.6}$$

Inequality (3.4) is associated with the operator identity (2.14).

Lemma 3.1. *Let a matrix-valued function $v(z)$ satisfying inequality* (3.1) *and the condition*

$$\det[u_1(-iv(z))+u_2]\neq 0 \tag{3.7}$$

be given. Then the matrix-valued function

$$\tilde{v}(z)=i[u_2(-iv(z))+u_1]\,[u_1(-iv(z))+u_2]^{-1} \tag{3.8}$$

satisfies inequality (3.4).

Proof. Recalling (2.12), we rewrite (3.5) in the form

$$\widetilde{B}(z)=(I-Az)^{-1}[\Phi_1\ \ \Phi_2]U\begin{bmatrix} I \\ -i\tilde{v}(z)\end{bmatrix}$$

i.e.,

$$\widetilde{B}(z)=(I-Az)^{-1}[\Phi_1\ \ \Phi_2]\begin{bmatrix} u_2-iu_1\tilde{v}(z) \\ u_1-iu_2\tilde{v}(z)\end{bmatrix}. \tag{3.9}$$

It follows from formulas (3.2), (3.8), and (3.9) that

$$\widetilde{B}(z)=B(z)[-iu_1v(z)+u_2]^{-1}. \tag{3.10}$$

Next, (3.8) yields

$$\widetilde{C}(z)=[-iu_1v(z)+u_2]^{*-1}C(z)[-iu_1v(z)+u_2]^{-1}. \tag{3.11}$$

Finally, by (3.1), (3.4), (3.10) and (3.11) we have

$$\widetilde{L}(z)=\begin{bmatrix} I & 0 \\ 0 & [-iu_1v(z)+u_2]^{*-1}\end{bmatrix}L(z)\begin{bmatrix} I & 0 \\ 0 & [-iu_1v(z)+u_2]^{-1}\end{bmatrix}.$$

Hence, $L(z)\geq 0$ implies that $\widetilde{L}(z)\geq 0$, as claimed. □

5.3.2. Let the spectral data $\tau(u)$ and α of a system (1.3) be given. Then using formulas (1.26) and (3.8) we find the matrix-valued function $v(z)$ of the corresponding system. Assuming that (3.7) holds, we can next obtain $\tilde{v}(z)$. The matrix-valued function $\tilde{v}(z)$ belongs to the Nevanlinna class and hence admits the representation

$$\tilde{v}(z)=\tilde{\gamma}z+\tilde{\alpha}+\int_{-\infty}^{\infty}\left(\frac{1}{u-z}-\frac{u}{1+u^2}\right)d\tilde{\tau}(u),\qquad \tilde{\gamma}\geq 0,\qquad \tilde{\alpha}=\tilde{\alpha}^*. \tag{3.12}$$

We will assume that

$$\tilde{\gamma}=0. \tag{3.13}$$

The next step is to construct the operators

$$S=\int_{-\infty}^{\infty}(I-Au)^{-1}\Phi_2\,[d\tilde{\tau}(u)]\,\widetilde{\Phi}_2^*(I-A^*u)^{-1} \tag{3.14}$$

and

$$\widetilde{\Phi}_1=-i\int_{-\infty}^{\infty}\left[A(I-Au)^{-1}+\frac{u}{1+u^2}I\right]\widetilde{\Phi}_2 d\tilde{\tau}(u)+i\widetilde{\Phi}_2\tilde{\alpha}. \tag{3.15}$$

Summing up, we have the following recipe for solving the inverse problem on the line $(-\infty,\infty)$ for the system (1.3).

Theorem 3.1. *Let $\tau(u)$ be a monotonically increasing $2m \times 2m$ matrix-valued function and let α be a constant selfadjoint $2m \times 2m$ matrix. Let the operators $A_{\pm}$, $\Phi_2^{\pm}$ be fixed and satisfy the following conditions:*

1) *In the spaces $\mathfrak{H}_+$ and $\mathfrak{H}_-$ there are continuous monotonically increasing families of orthogonal projectors P_ξ^+, P_ξ^- such that*

$$A_{\pm}^* P_\xi^{\pm} = P_\xi^{\pm} A_{\pm}^* P_\xi^{\pm}, \qquad P_0 = 0, \qquad P_l^{\pm} = I_{\pm}.$$

2) *The spectra of the operators A_+ and A_- are concentrated in zero and there exists a constant $M > 0$ such that*

$$\|(P_{\xi+\Delta\xi}^{\pm} - P_\xi^{\pm})A_{\pm}(P_{\xi+\Delta\xi}^{\pm} - P_\xi^{\pm})\| \le M|\Delta\xi|, \qquad 0 \le \xi \le l.$$

3) *Zero is not an eigenvalue of operator*

$$A = \begin{bmatrix} A_+ & 0 \\ 0 & A_- \end{bmatrix}.$$

Assume further that the relations (3.7), (3.13) *are satisfied and that the operators S and $\widetilde{\Phi}_1$ defined by relations* (3.14) *and* (3.15) *are bounded and have the form*

$$S = \begin{bmatrix} S_+ & 0 \\ 0 & S_- \end{bmatrix}, \qquad \widetilde{\Phi}_1 = \frac{1}{\sqrt{2}} \begin{bmatrix} \Phi_1^+ & \Phi_1^+ \\ \Phi_1^- & -\Phi_1^- \end{bmatrix}. \tag{3.16}$$

If the operators S_+ and S_- are positive, then the system

$$\mathcal{W}(x,z) = I_{2m} + izJ \int_0^x [dB(t)]\mathcal{W}(t,z), \qquad -l \le x \le l,$$

with the given spectral data $\tau(u)$ and α can be found by means of the formula (2.8) *and the relations*

$$B(x) = TB_0(x)T^{-1} = \operatorname{diag}\{B_+(x), B_-(x)\}, \tag{3.17}$$

$$B(\pm x) = \pm B_{\pm}(x), \quad x \ge 0. \tag{3.18}$$

Proof. Given the spectral data $\tau(u)$ and α, construct the function $v(z)$ by means of formula (1.26). Next, using formula (3.8) find $\tilde{v}(z)$. The matrix-valued function $\tilde{v}(z)$ satisfies inequality (3.4) (see Lemma 3.1), where the operators S and $\widetilde{\Phi}_1$ are defined by relations (3.14) and (3.15), respectively. Then the representations

$$S = \int_{-\infty}^{\infty} (I - Au)^{-1}\Phi_2 \,[d\tau(u)]\, \Phi_2^*(I - A^*u)^{-1}, \tag{3.19}$$

and

$$\Phi_1 = -i\int_{-\infty}^{\infty} \left[A(I - Au)^{-1} + \frac{u}{1+u^2}I\right] \Phi_2 d\tau(u) + i\Phi_2\alpha \tag{3.20}$$

hold. This shows that $\tau(u)$ and α are spectral data of the system (1.14) (Theorem 2.6, Chapter 4). From conditions (3.16) we conclude that $B(x)$ is diagonal. Then from formulas (3.17) and (3.18) we obtain a system (1.3) on the segment $[-l, l]$. By Theorem 1.1, the spectral data of the system (1.3) coincide with the given $\tau(u)$ and α. This completes the proof of the theorem. □

Chapter 6
Weyl-Titchmarsh Functions of Periodic Canonical Systems

Let $\mathcal{H}(x)$ be a $n \times n$ $(n = 2m)$ matrix-valued function whose elements belong to $L_1(0, l)$ and which satisfies

$$\mathcal{H}(x) \geq 0, \qquad \int_0^l \mathcal{H}(x)dx > 0, \qquad \mathcal{H}(x + l) = \mathcal{H}(x). \tag{0.1}$$

Let J denote the matrix

$$J = \begin{bmatrix} 0 & I_m \\ I_m & 0 \end{bmatrix}. \tag{0.2}$$

The corresponding canonical system of differential equations has the form

$$\frac{dY}{dx} = izJ\mathcal{H}(x)Y, \qquad -\infty < x < \infty, \tag{0.3}$$

where $Y(x, z)$ is an $n \times 1$ column vector. Denote by $\mathcal{W}(x, z)$ the matrix solution of system (0.3) satisfying the normalization condition

$$\mathcal{W}(0, z) = I_n. \tag{0.4}$$

In Chapter 5 we introduced the analogue of the Weyl-Titchmarsh function $v(z)$ for a system of the form (0.3) and clarified the connection between $v(z)$ and the spectral characteristics of the system (0.3). Therein we also gave a recipe for solving the inverse problem under certain additional conditions, i.e., a recipe for reconstructing $\mathcal{H}(x)$ from the known matrix function $v(z)$. In the present chapter we investigate the Weyl-Titchmarsh matrix function $v(z)$ in the periodic case.

6.1. Multipliers and their behavior

6.1.1. This section is based on a celebrated article by M. G. Kreĭn and G. Ya. Lyubarskiĭ [52].

The eigenvalues $\rho_k(z)$ of the monodromy matrix $\mathcal{W}(z)$ are called the *multipliers* of the system (0.3). Denote by $h_k(z)$ the corresponding eigenvectors and by $q_k(z)$ the order of the corresponding Jordan cell.

The coefficients of the polynomial

$$D(\rho) = D(\rho, z) = \det\left[W(z) - \rho I_{2m}\right]$$

are entire functions of z. The equation

$$D(\rho, z) = 0 \tag{1.1}$$

defines ρ as a multi-valued analytic function of z. If for $z = z_0$ $(-\infty < z_0 < \infty)$ one of the multipliers is equal to ρ_0, then using a Newton diagram it is possible to recursively determine the terms of the expansions of those branches $\rho(z)$ which take the value ρ_0 in $z = z_0$. If the number of these branches is denoted by d, and the branches themselves are denoted by $\rho_k(\lambda)$, $k = 1, 2, \ldots, d$, then the aforementioned expansions have the form

$$\rho_k(z) = \rho_{k,0} + \sum_{s=1}^{\infty} c_{k,s}(z - z_0)^{s/q_k}. \tag{1.2}$$

It follows from (1.2) that the corresponding eigenvectors $h_k(z)$ can be represented in the form

$$h_k(z) = \sum_{s=0}^{\infty} h_{k,s}(z - z_0)^{s/q_k}. \tag{1.3}$$

Let us also expand $W(z)$ in a series

$$W(z) = W_0 + \sum_{s=1}^{\infty} W_s(z - z_0)^s, \qquad W_0 = W(z_0). \tag{1.4}$$

From equalities (1.1)–(1.4) and the relation

$$W(z)h_k(z) = \rho_k(z)h_k(z) \tag{1.5}$$

we deduce that

$$W_0 h_{k,s} = \rho_{k,0}h_{k,s} + c_{k,s}h_{k,0} + c_{k,s-1}h_{k,1} + \cdots + c_{k,1}h_{k,s-1}, \tag{1.6}$$

where $1 \le s \le q_k - 1$. This shows that $h_{k,0}$ is an eigenvector corresponding to the multiplier $\rho_{k,0} = \rho_k(z_0)$ and that $h_{k,s}$, $1 \le s \le q_k - 1$ are corresponding root vectors.

Lemma 1.1. *Let conditions* (0.1)–(0.4) *be satisfied. Then*

$$(JW_1 h_{k,0}, h_{k,0}) \neq 0. \tag{1.7}$$

Proof. Let $\mathcal{W}(x, z)$ be a solution of the system (0.3), (0.4). Substituting the expression

$$\mathcal{W}(x, z) = \mathcal{W}_0(x, z_0) + \mathcal{W}_1(x, z_0)(z - z_0) + \cdots \tag{1.8}$$

in (0.3) and equating the coefficients of $z - z_0$ in the left- and right-hand sides, we obtain

$$\frac{d\mathcal{W}_1(x, z_0)}{dx} = iJ\mathcal{H}(x)\mathcal{W}_0(x, z_0) + iz_0 J\mathcal{H}(x)\mathcal{W}_1(x, z_0), \qquad \mathcal{W}_1(0, z_0) = 0. \tag{1.9}$$

Integrating the system (1.9) we have

$$\mathcal{W}_1(x,z_0) = i\mathcal{W}_0(x,z_0)\int_0^x \mathcal{W}_0^{-1}(t,z_0)J\mathcal{H}(t)\mathcal{W}_0(t,z_0)\,dt. \tag{1.10}$$

In particular, setting here $x = l$ yields

$$W_1 = \mathcal{W}_1(l,z_0) = iW_0\int_0^l \mathcal{W}_0^{-1}(t,z_0)J\mathcal{H}(t)\mathcal{W}_0(t,z_0)\,dt. \tag{1.11}$$

Using relation (1.11) and the equality

$$\mathcal{W}_0^*(x,z_0)J\mathcal{W}_0(x,z_0) = J, \qquad z_0 = \bar{z}_0, \tag{1.12}$$

we calculate the inner product

$$\big(JW_1h_{k,0},h_{r,0}\big) = i\int_0^l \big(\mathcal{W}_0^{-1}(t,z_0)J\mathcal{H}(t)x_k(t),W_0^{-1}h_{r,0}\big)\,dt, \tag{1.13}$$

where

$$x_k(t) = \mathcal{W}_0(t,z_0)h_{k,0}. \tag{1.14}$$

Relations (1.6) and (1.12)–(1.14) yield the equality

$$\big(JW_1h_{k,0},h_{r,0}\big) = \frac{i}{\bar{\rho}_{k,0}}\int_0^l \big(\mathcal{H}(t)x_k(t),x_r(t)\big)\,dt. \tag{1.15}$$

By (0.1),

$$\int_0^l \big(\mathcal{H}(t)x_k(t),x_k(t)\big)\,dt \geq 0. \tag{1.16}$$

If equality holds in (1.16), then

$$\mathcal{H}(t)x_k(t) = 0, \tag{1.17}$$

which in conjunction with (1.14) yields

$$\frac{dx_k(t)}{dt} = iz_0J\mathcal{H}(t)x_k(t) = 0.$$

This means that

$$x_k(t) = x_k(0) \neq 0.$$

However, by (0.1),

$$\int_0^l \big(\mathcal{H}(t)x_k(0),x_k(0)\big)\,dt > 0,$$

and so the assumption that equality holds in (1.16) leads to a contradiction. This completes the proof of the lemma. □

Now let us study the behavior of the vectors $h_{k,p}$, following the paper [52].

Theorem 1.1. *Suppose that relations* (0.1)–(0.4) *are satisfied. Then*

$$(Jh_{k,p}, h_{k,s}) = 0, \qquad p + s < q_k - 1, \tag{1.18}$$

$$(Jh_{k,p}, h_{k,s}) = \gamma_k^s c_k, \qquad p + s = q_k - 1, \tag{1.19}$$

where

$$c_k \neq 0, \qquad c_{k,1} \neq 0, \tag{1.20}$$

are given by

$$c_k = \frac{(JW_1 h_{k,0}, h_{k,0})}{c_{k,1}}, \tag{1.21}$$

$$\gamma_k = -\frac{\rho_{k,0}\bar{c}_{k,1}}{\bar{\rho}_{k,0}c_{k,1}} = e^{i\alpha_k}. \tag{1.22}$$

Proof. To simplify notations, the index k will be omitted. By (1.12),

$$(JW_0 h_p, W_0 h_s) = (Jh_p, h_s). \tag{1.23}$$

On the other hand, by (1.6) we have

$$\begin{gathered}(JW_0 h_p, W_0 h_s) = \\ = \big(J(\rho_0 h_p + c_1 h_{p-1} + \cdots + c_p h_0), (\rho h_s + c_1 h_{s-1} + \cdots + c_s h_0)\big).\end{gathered} \tag{1.24}$$

Comparing (1.23) and (1.24), we conclude that

$$\begin{gathered}\big(J(\rho_0 h_p + c_1 h_{p-1} + \cdots + c_p h_0), (c_1 h_{s-1} + \cdots + c_s h_0)\big) + \\ + \big(J(c_1 h_{p-1} + \cdots + c_p h_0), \rho_0 h_s\big) = 0.\end{gathered} \tag{1.25}$$

Setting here $s = 0$, we get

$$(Jh_p, h_0) = 0, \qquad 0 \leq p \leq q - r. \tag{1.26}$$

Next, taking $s = 1, 2, \ldots, q - r$ and using equalities (1.25) and (1.26) we obtain the relations (1.18). Using again formulas (1.2)–(1.4) and equality (1.5) we obtain

$$W_1 h_0 + W_0 h_q = c_q h_0 + c_{q-1} h_1 + \cdots + c_1 h_{q-1} + \rho_0 h_q. \tag{1.27}$$

Analogously to (1.26), we derive the relation

$$\big(J(c_q h_0 + c_{q-1} h_1 + \cdots + c_1 h_{q-1} - W h_0), h_0\big) = 0, \tag{1.28}$$

which in conjunction with (1.18) yields

$$c_1 (Jh_{q-1}, h_0) = (JW_1 h_0, h_0). \tag{1.29}$$

Finally, (1.29) and Lemma 1 show that

$$c_k \neq 0. \tag{1.30}$$

Now let $s > 0$ and $p + s = q$. Then, by (1.18) and (1.25),

$$\rho_0 \bar{c}_1 (Jh_p, h_{s-1}) + c_1 \bar{\rho}_0 (Jh_p, h_{s-1}) = 0,$$

i.e.,

$$(Jh_{p-1}, h_s) = \gamma_k (Jh_p, h_{s-1}), \qquad p + s = q, \tag{1.31}$$

where the multiplier γ_k is defined by relation (1.22). Relation (1.19) is a direct consequence of (1.29) and (1.31). Finally, (1.20) follows from (1.18) and (1.29). The theorem is proved. □

6.1.2. Now let us consider the case when

$$z_0 = \bar{z}_0, \qquad \rho_{k,0} = \rho_{r,0}, \qquad k \neq r, \qquad |\rho_{k,0}| = 1. \tag{1.32}$$

With no loss of generality we may assume that

$$(JW_1 h_{k,0}, h_{r,0}) = 0, \qquad k \neq r. \tag{1.33}$$

The following assertion can be proved in much the same manner as Theorem 1.1.

Theorem 1.2. *Suppose that the relations* (0.1)–(0.4) *and* (1.32), (1.33) *are satisfied. Then*

$$(Jh_{k,p}, h_{r,s}) = 0, \qquad p + s \leq q_k - 1, \quad 0 \leq p \leq q_k - 1 \quad 0 \leq s \leq q_r - 1. \tag{1.34}$$

Remark 1.1. Theorems 1.1 and 1.2 were proved in [52] under the assumption that $s = 0$.

6.1.3. Here we consider the case where

$$z_0 = \bar{z}_0, \qquad \rho_{k,0} \bar{\rho}_{r,0} \neq 1. \tag{1.35}$$

Theorem 1.3. *Suppose that the relations* (0.1)–(0.4) *and* (1.35) *hold. Then*

$$(Jh_{k,p}, h_{r,s}) = 0, \qquad 0 \leq p \leq q_k - 1, \qquad 0 \leq s \leq q_r - 1. \tag{1.36}$$

Proof. Since $\rho_{k,0}$ and $\rho_{r,0}$ are eigenvalues of the J-unitary matrix W_0 it follows from the inequality $\rho_{k,0} \bar{\rho}_{r,0} \neq 1$ that the corresponding root subspaces are J-orthogonal, i.e., equalities (1.36) hold. The theorem is proved. □

6.2. Weyl-Titchmarsh functions

6.2.1. Let r be a positive integer. Let us introduce the matrix-valued functions

$$a_r(z) = J[W^r(z) + W^{-r}(z)]J, \tag{2.1}$$

$$b_r(z) = -J[W^r(z) - W^{-r}(z)]J. \tag{2.2}$$

Consider the linear-fractional transformation

$$v_r(z) = iJ[a_r(z)J\mathcal{P}(z) + b_r(z)\mathcal{Q}(z)]\,[b_r(z)J\mathcal{P}(z) + a_r(z)\mathcal{Q}(z)]^{-1}, \tag{2.3}$$

where $\mathcal{P}(z)$, $\mathcal{Q}(z)$ is a pair of meromorphic matrix-valued functions such that

$$\det[\mathcal{P}^*(z)\mathcal{P}(z) + \mathcal{Q}^*(z)\mathcal{Q}(z)] \not\equiv 0, \quad \text{Im } z > 0, \tag{2.4}$$

$$\mathcal{P}^*(z)\mathcal{Q}(z) + \mathcal{Q}^*(z)\mathcal{P}(z)] \geq 0, \quad \text{Im } z > 0. \tag{2.5}$$

The functions $v_r(z)$ form a *Weyl disc*. Moreover, there exists the limit (see Theorem 2.1)

$$v(z) = \lim_{r\to\infty} v_r(z), \quad \text{Im } z > 0, \tag{2.6}$$

which does not depend on the choice of $\mathcal{P}(z)$ and $\mathcal{Q}(z)$. The matrix-valued function $v(z)$ is called the *Weyl-Titchmarsh function* of the system (0.3). The connection between $v(z)$ and the spectral characteristics of the system (0.3) was elucidated in Chapter 5.

Proposition 2.1. *The matrix-valued function $W(z)$ has the following properties:*

$$W^*(z)JW(z) < J, \quad \text{Im}\, z > 0, \tag{2.7}$$

$$W^*(z)JW(z) > J, \quad \text{Im}\, z < 0, \tag{2.8}$$

$$W^*(z)JW(z) = J, \quad \text{Im}\, z = 0, \tag{2.9}$$

Proof. Since $\mathcal{W}(x,z)$ is a matrix solution of system (0.3), we have

$$\frac{d}{dx}[\mathcal{W}^*(x,z)J\mathcal{W}(x,z)] = \frac{\bar{z}-z}{i}\mathcal{W}^*(x,z)\mathcal{H}(x)\mathcal{W}(x,z).$$

Taking into account condition (0.4), we obtain

$$\mathcal{W}^*(x,z)J\mathcal{W}(x,z) = J - \frac{z-\bar{z}}{i}\int_0^x \mathcal{W}^*(t,z)\mathcal{H}(t)\mathcal{W}(t,z)\,dt. \tag{2.10}$$

Setting here $x = l$ yields

$$W^*(z)JW(z) = J - \frac{z-\bar{z}}{i}\int_0^l \mathcal{W}^*(t,z)\mathcal{H}(t)\mathcal{W}(t,z)\,dt. \tag{2.11}$$

We claim that

$$\int_0^l \mathcal{W}^*(t,z)\mathcal{H}(t)\mathcal{W}(t,z)\,dt > 0. \tag{2.12}$$

Indeed, suppose that (2.12) is not true. Then one can find a column vector h such that

$$\int_0^l Y^*(t,z)\mathcal{H}(t)Y(t,z)\,dt = 0, \tag{2.13}$$

where $Y(t,z) = \mathcal{W}(t,z)h$. Then the equality

$$Y^*(t,z)\mathcal{H}(t)Y(t,z) = 0$$

holds almost everywhere, i.e.,

$$\mathcal{H}(t)Y(t,z) = 0.$$

Using (0.3), we conclude that

$$Y(t,z) = Y(0,z) = h.$$

But then (2.13) contradicts (0.1). Relations (2.7)–(2.9) are immediate consequences of (2.10) and (2.12). The proposition is proved. □

Thus, the matrix $W(z)$ is strictly J-contractive in the upper half-plane $\operatorname{Im} z > 0$. As V. P. Potapov showed ([65], Ch.2, §2), strictly J-contractive matrices admit a representation

$$W(z) = U(z)\begin{bmatrix} D_1(z) & 0 \\ 0 & D_2(z) \end{bmatrix} U^{-1}(z), \tag{2.14}$$

where $D_1(z)$ and $D_2(z)$ are $m \times m$ matrices and

$$\|D_1(z)\| < 1, \qquad \|D_2^{-1}(z)\| < 1, \qquad \operatorname{Im} z > 0. \tag{2.15}$$

Theorem 2.1. *Let a periodic canonical system* (0.1)–(0.3) *be given. Then the limit* (2.6) *exists and*

$$v(z) = iU(z)jU^{-1}(z)J, \qquad \operatorname{Im} z > 0, \tag{2.16}$$

where

$$j = \begin{bmatrix} I_m & 0 \\ 0 & -I_m \end{bmatrix}. \tag{2.17}$$

Proof. Let us rewrite formula (2.3) in the form

$$v_r(z) = iJb_r(z)[\gamma_r(z)J\mathcal{P}(z) + \mathcal{Q}(z)]\,[\gamma_r^{-1}(z)J\mathcal{P}(z) + \mathcal{Q}(z)]^{-1}, \tag{2.18}$$

where

$$\gamma_r(z) = b_r^{-1}(z)a_r(z). \tag{2.19}$$

The invertibility of the matrices $a_r(z)$ and $b_r(z)$ follows from formulas (2.1), (2.2) and (2.14). Next, using (2.1), (2.2) and (2.14) one more time, we conclude that

$$\gamma_r(z) = -JU(z)\begin{bmatrix} A_1(r,z) & 0 \\ 0 & A_2(r,z)\end{bmatrix} U^{-1}(z)J, \tag{2.20}$$

where

$$A_k(r,z) = [D_k^r(z) - D_k^{-r}(z)]^{-1}\,[D_k^r(z) + D_k^{-r}(z)]. \tag{2.21}$$

Finally, (2.15) and (2.21) guarantee that

$$\gamma(z) = \lim_{r\to\infty} \gamma_r(z) = JU(z)jU^{-1}(z)J, \tag{2.22}$$

i.e.,

$$\gamma(z) = \gamma^{-1}(z). \tag{2.23}$$

Formula (2.16) is now seen to be a consequence of (2.18), (2.22) and (2.23). □

6.2.2. Let us prove that the matrix-valued function $v(z)$ admits an analytic continuation.

Theorem 2.2. *The matrix-valued function $v(z)$ admits an analytic continuation to the whole complex plane with cuts along the rays $(-\infty, \lambda_1]$ and $[\lambda_2, \infty)$ of the real line, where $\lambda_1 < 0 < \lambda_2$.*

Proof. Denote by $\rho_k(z)$ the multipliers of the system (0.1)–(0.3) and by $h_k(z)$ the corresponding eigenvectors. As is known (see [52]), there exist numbers λ_1 and λ_2, $\lambda_1 < 0 < \lambda_2$, such that for z in the interval (λ_1, λ_2) the matrix $W(z)$ does not have Jordan cells and $|\rho_k(z)| = 1$ $(1 \le k \le 2m)$. This means that when $z \in (\lambda_1, \lambda_2)$ there are m *multipliers of the first kind* $\rho_k(z)$ $(1 \le k \le m)$ and m *multipliers of the second kind* $\rho_k(z)$ $(m+1 \le k \le 2m)$, i.e.,

$$\begin{aligned} (Jh_k(z), h_k(z)) &> 0, \quad 1 \le k \le m, \\ (Jh_k(z), h_k(z)) &< 0, \quad m+1 \le k \le 2m. \end{aligned} \tag{2.24}$$

Set

$$D_1(z) = \operatorname{diag}\{\rho_1(z), \cdots, \rho_m(z)\}, \qquad D_2(z) = \operatorname{diag}\{\rho_{m+1}(z), \cdots, \rho_{2m}(z)\}.$$

Now write the matrix-valued function $W(z)$ in the form (2.14). The matrix-valued functions $U(z)$, $U^{-1}(z)$, $D_1(z)$, $D_2(z)$ are analytic in some domain G containing the interval (λ_1, λ_2). Let $z \in G$ and $\operatorname{Im} z < 0$; then representation (2.10) remains valid. However, relations (2.15) are replaced by

$$\|D_1^{-1}(z)\| < 1, \qquad \|D_2(z)\| < 1, \qquad \operatorname{Im} z < 0, \qquad z \in G. \tag{2.25}$$

Since $W(z)$ does not have eigenvalues of modulus 1 when $\operatorname{Im} z < 0$, the matrix-valued functions $U(z)$, $U^{-1}(z)$ satisfying relations (2.14) and (2.25) can be chosen to be analytic at any point of the boundary of the lower half-plane. This proves the theorem. □

6.2.3. Let us denote

$$B(z) = U^*(z)JU(z), \qquad z = \bar{z}. \tag{2.22}$$

Here we will consider the case when on a certain interval of the real line, $a < z < b$, all the multipliers lie outside the circle $|\zeta| = 1$, i.e.,

$$|\rho_k(z)| < 1, \quad 1 \le k \le m, \qquad |\rho_k(z)| > 1, \quad m < k \le 2m, \qquad a < z < b. \tag{2.27}$$

Denote by $L_1(z)$ and $L_2(z)$ the invariant subspaces of $W(z)$ corresponding to the eigenvalues $|\rho_k(z)| < 1$ and $|\rho_k(z)| > 1$, respectively. As we already know (Theorem 1.3),

$$(J\xi_k, \xi_k) = 0, \qquad \xi_k \in L_k \quad (k = 0, 1). \tag{2.28}$$

By (2.28), the matrix $B(z)$ has the form

$$B(z) = \begin{bmatrix} 0 & B_1(z) \\ B_1^*(z) & 0 \end{bmatrix}.$$

It follows from (2.26) that in the case under consideration equality (2.16) can be recast as

$$v(z) = U(z) \begin{bmatrix} 0 & iB_1^{*-1}(z) \\ -iB_1^{-1}(z) & 0 \end{bmatrix} U^*(z),$$

i.e.,

$$v(z) = v^*(z), \qquad a < z < b. \tag{2.30}$$

From (2.30) and the results of Chapter 5 we deduce the following assertion.

Theorem 2.3. *If all the multipliers of $W(z)$ lie outside the circle $|\zeta| = 1$ when $a < z < b$, then the interval (a, b) is a gap in the spectrum of the system* (0.1)–(0.3).

6.3. Singular points of the Weyl-Titchmarsh matrix function

In this section we study the singularities of the Weyl-Titchmarsh matrix-valued function $v(z)$ of the system (0.1)–(0.3).

6.3.1. Let us order the eigenvalues of $W(z_0)$ so that

$$|\rho_{k,0}| = 1, \quad 1 \le k \le N; \qquad |\rho_{k,0}| \ne 1, \quad k > N; \qquad (z_0 = \bar{z}_0). \tag{3.1}$$

Let us denote $\eta_{k,r}(z)$ $(1 \le k \le N)$ the different branches of the multi-valued function

$$\eta_{k,r}(z) = \sum_{s=0}^{\infty} h_{k,s} e^{i2\pi rs/q_k} (z - z_0)^{s/q_k}, \qquad 0 \le r \le q_k - 1, \tag{3.2}$$

where

$$z > z_0, \qquad \arg\,(z - z_0)^{s/q_k} = 0, \qquad q_k = q_k(z_0). \tag{3.3}$$

Then relations (1.18) and (1.19) yield the asymptotic equality

$$(J\eta_{k,u}(z), \eta_{k,v}(z)) = (z - z_0)^{(q_k-1)/q_k} \sum_{p+s=q_k-1} e^{i2\pi(up-vs)/q_k} e^{i\alpha_k s} c_k + \\ + O(|z - z_0|), \qquad z \to z_0, \tag{3.4}$$

where

$$c_k = \frac{(JW_1 h_{k,0}, h_{k,0})}{c_{k,1}} \neq 0. \tag{3.5}$$

We are interested in the vectors $\eta_{k,r}(z)$ $(1 \le k \le N)$ when $|z - z_0| < \varepsilon$ and $\operatorname{Im} z > 0$. Denote by q_k^+ and q_k^- the number of the vectors $\eta_{k,r}(z)$ of the first and second kind, respectively. By the Poincaré–Lyapunov theorem asserting that the multipliers are symmetrically positioned with respect to the circle $|z - z_0| = 1$, we have

$$\sum_{k=1}^{N} q_k^+ = \sum_{k=1}^{N} q_k^- = Q_1. \tag{3.6}$$

Here the number Q_1 does not depend on z provided that ε is small. Further, denote by $L^+(z)$ and $L^-(z)$ the subspaces with bases provided by the vectors $\eta_{k,r}^+(z)$ and $\eta_{k,r}^-(z)$ $(k > N)$ for which $|\rho_{k,0}| < 1$ and $|\rho_{k,0}| > 1$, respectively. A second application of the Poincaré–Lyapunov theorem yields

$$\dim L^+(z) = \dim L^-(z) = Q_2 = m - Q_1. \tag{3.7}$$

The chosen bases degenerate when $q_k > 1$. However, by the Riesz spectral theorem ([70], Ch. 11), the subspaces $L^+(z)$ and $L^-(z)$ admit bases of analytic vector-valued functions $\widetilde{h}_k(z)$ $(Q_1 < k \le m)$ and $\widetilde{h}_k(z)$ $(m + Q_1 < k \le 2m)$ which do not degenerate. Let us partition the vectors $\eta_{k,r}(z)$ $(1 \le k \le N)$ into two subsets: the vectors $h_1^+(z), h_2^+(z), \dots, h_{Q_1}^+(z)$ of the first kind and the vectors $h_1^-(z), h_2^-(z), \dots, h_{Q_1}^-(z)$ of the second kind. Consider the $2m \times 2m$ matrix

$$\widetilde{U}(z) = [h_1^+(z)\ h_2^+(z)\ \cdots\ h_m^+(z)\ h_1^-(z)\ h_2^-(z)\ \cdots\ h_m^-(z)], \tag{3.8}$$

where

$$h_k^+(z) = \widetilde{h}_k(z), \qquad Q_1 < k \le m, \tag{3.9}$$

$$h_k^-(z) = \widetilde{h}_{k+m}(z), \qquad Q_1 < k \le m. \tag{3.10}$$

With this notation, one has the relation

$$W(z) = \widetilde{U}(z) \begin{bmatrix} \widetilde{D}_1(z) & 0 \\ 0 & \widetilde{D}_2(z) \end{bmatrix} \widetilde{U}^{-1}(z). \tag{3.11}$$

Comparing (2.14) and (3.11) we obtain

$$\widetilde{U}(z) = U(z)R(z), \tag{3.12}$$

where the matrix $R(z)$ has following block structure:

$$R(z) = \begin{bmatrix} r_1(z) & 0 \\ 0 & r_2(z) \end{bmatrix}. \tag{3.13}$$

By (3.11)–(3.13), formula (2.16) can be rewritten as

$$v(z) = i\widetilde{U}(z)j\widetilde{U}^{-1}(z)J, \qquad \operatorname{Im} z > 0. \tag{3.14}$$

Now let us introduce a new matrix $\widehat{U}(z)$, defined as

$$\widehat{U}(z) = [U_1(z)\ U_2(z)\ \cdots\ U_{N+2}(z)], \tag{3.15}$$

where the matrices $U_k(z)$ have the form

$$U_k(z) = [\eta_{k,0}(z)\ \eta_{k,1}(z)\ \cdots\ \eta_{k,q_k-1}(z)], \qquad 1 \le k \le N, \tag{3.16}$$

$$U_{N+1}(z) = [h^+_{Q_1+1}(z)\ \cdots\ h^+_m(z)] \tag{3.17}$$

$$U_{N+2}(z) = [h^-_{Q_1+1}(z)\ \cdots\ h^-_m(z)]. \tag{3.18}$$

It is readily verified that the matrix $\widehat{U}(z)$ is obtained from the matrix $\widetilde{U}(z)$ by permuting the columns. Hence, equality (3.14) can be written in the form

$$v(z) = i\widehat{U}(z)\widehat{j}\widehat{U}^{-1}(z)J, \qquad \operatorname{Im} z > 0, \tag{3.19}$$

where

$$\widehat{j} = \operatorname{diag}\{\varepsilon_1, \varepsilon_2, \cdots, \varepsilon_{2m}\}, \qquad \varepsilon_r = \pm 1.$$

Here $\varepsilon_r = 1$ if the column of $\widehat{U}(z)$ is an eigenvector of the first kind and $\varepsilon_r = -1$ if the corresponding column is an eigenvector of the second kind. We remark that equality (3.19) is valid only in some neighbourhood of z_0. The matrices $U_k(z)$ have the following form:

$$U_k(z_0) = [\underbrace{\eta_{k,0}\ \eta_{k,0}\ \cdots\ \eta_{k,0}}_{q_k}] \tag{3.20}$$

when $z = z_0$ and $1 \le k \le N$.

Further we shall assume that q_k $(1 \le k \le N)$ are arranged in decreasing order and that q_1 is repeated N_1 times. We are now ready to formulate the main result of this section.

Theorem 3.1. *The Weyl-Titchmarsh matrix function $v(z)$ of the canonical system (0.1)–(0.3) admits the asymptotic representation*

$$v(z) = (z - z_0)^{-(q_1-1)/q_1}[A + o(1)], \qquad z > z_0, \tag{3.21}$$

in a neighbourhood of any point z_0, $z_0 = \bar{z}_0$. Here A is a $2m \times 2m$ constant matrix and

$$\operatorname{rank} A = N_1. \tag{3.22}$$

6.3.2. In order to prove Theorem 3.1 we need some additional facts.

First let us assume that

$$e^{i\alpha_k q_k} \neq 1. \tag{3.23}$$

Then

$$\sum_{p+s=q_k-1} e^{i2\pi(up-vs)/q_k} e^{i\alpha_k s} = s_{u,v}, \tag{3.24}$$

where we denote

$$s_{u,v} = \frac{1 - e^{i\alpha_k q_k}}{e^{i2\pi u/q_k} - e^{-i2\pi v/q_k} e^{i\alpha_k}}. \tag{3.25}$$

From relations (3.4) and (3.24) we deduce that

$$(J\eta_{k,u}(z), \eta_{k,v}(z)) = (z - z_0)^{(q_k-1)/q_k} c_k s_{u,v} + O(|z - z_0|). \tag{3.26}$$

Let us introduce the matrix

$$S = \{s_{v,u}\}_{v,u=0}^{q_k-1} = \beta_k \left\{ \frac{1}{a_v + b_u} \right\}_{v,u=0}^{q_k-1}, \tag{3.27}$$

where

$$a_v = -e^{-i2\pi v/q_k} e^{i\alpha_k}, \qquad b_u = e^{i2\pi u/q_k}, \qquad \beta_k = 1 - e^{i\alpha_k q_k}. \tag{3.28}$$

Lemma 3.1. *Suppose that condition (3.23) is satisfied. Then the matrix S given by (3.27) is invertible and its inverse $T = S^{-1}$ has the form*

$$T = \{t_{p,r}\}_{p,r=0}^{q_k-1}, \tag{3.29}$$

where

$$t_{p,r} = \frac{\beta_k}{q_k^2} \frac{e^{-i\alpha_k(q_k-1)}}{e^{i2\pi r/q_k} - e^{-i2\pi p/q_k} e^{i\alpha_k}}. \tag{3.30}$$

Proof. We will use the well-known Cauchy formula (see [64]):

$$\Delta = \det \left\{ \frac{1}{a_v + b_u} \right\}_{v,u=0}^{q-1} = \frac{\prod\limits_{q-1 \geq j > p \geq 0} (a_j - a_p)(b_j - b_p)}{\prod\limits_{v,u=0}^{q-1} (a_v + b_u)}. \tag{3.31}$$

Denote by $A_{p,r}$ the cofactor of the p,r-element of this Cauchy determinant. Then, by (3.31),

$$\frac{A_{p,r}}{\Delta} = \frac{\prod_{v=0}^{q_k-1}(a_v+b_r)\prod_{u=0}^{q_k-1}(a_p+b_u)}{(a_p+b_r)\prod_{j\neq p}(a_j-a_p)\prod_{j\neq r}(b_j-b_r)}. \tag{3.32}$$

Recalling (3.28), we have

$$\prod_{j\neq p}(a_j-a_p) = e^{i2\pi p/q_k}e^{i\alpha_k(q_k-1)}\prod_{s=1}^{q_k-1}(1-e^{i2\pi s/q_k}), \tag{3.33}$$

$$\prod_{j\neq r}(b_j-b_r) = e^{-i2\pi r/q_k}\prod_{s=1}^{q_k-1}(e^{i2\pi s/q_k}-1). \tag{3.34}$$

Similarly, one can show that

$$\prod_{v=0}^{q_k-1}(a_v+b_r) = \prod_{s=0}^{q_k-1}(1-e^{i\alpha_k}e^{-i2\pi s/q_k}), \tag{3.35}$$

$$\prod_{u=0}^{q_k-1}(a_p+b_u) = e^{i\alpha_k q_k}\prod_{s=0}^{q_k-1}(e^{i2\pi s/q_k}e^{-i\alpha_k}-1). \tag{3.36}$$

Further, the equality

$$\prod_{s=0}^{q-1}(z-e^{-i2\pi s/q}) = z^q-1$$

yields the relations

$$\prod_{s=1}^{q_k-1}(1-e^{-i2\pi s/q_k}) = q_k, \tag{3.37}$$

$$\prod_{s=0}^{q_k-1}(e^{-i\alpha_k}-e^{-i2\pi s/q_k}) = e^{-i\alpha_k q}-1. \tag{3.38}$$

Using (3.37), we can recast the formulas (3.33) and (3.34) in the form

$$\prod_{j\neq p}(a_j-a_p) = e^{i2\pi p/q_k}\,e^{i\alpha_k(q_k-1)}\,q_k \tag{3.39}$$

and

$$\prod_{j\neq r}(b_j-b_r) = (-1)^{q_k-1}\,e^{-i2\pi r/q_k}\,q_k, \tag{3.40}$$

respectively. In the notations (3.28), formulas (3.35) and (3.36) become

$$\prod_{v=0}^{q_k-1}(a_v+b_r)=\beta_k,\qquad \prod_{u=0}^{q_k-1}(a_p+b_u)=(-1)^{q_k-1}\,\beta_k. \tag{3.41}$$

Substituting (3.39)–(3.41) in (3.32) we obtain the equality

$$\frac{A_{p,r}}{\Delta}=\frac{\beta_k^2}{q_k^2}\,\frac{e^{-i\alpha_k(q_k-1)}}{e^{i2\pi p/q_k}-e^{-i2\pi r/q_k}e^{i\alpha_k}}, \tag{3.42}$$

which in turn yields (3.30), as needed. □

Lemma 3.2. *The following identity holds:*

$$\sum_{p=0}^{q-1}(e^{i2\pi r/q}-e^{-i2\pi p/q}\,e^{i\alpha})^{-1}=\frac{e^{-i2\pi r/q}\,q}{1-e^{i\alpha q}}. \tag{3.43}$$

Proof. Let us write the left-hand side of (3.43) in the form

$$\frac{Q_r(z)}{\prod_{p=0}^{q-1}(e^{i2\pi r/q}-e^{-i2\pi p/q}\,e^{i\alpha})}, \tag{3.44}$$

where

$$Q_r(z)=\sum_{p=0}^{q-1}\prod_{s\neq p}(e^{i2\pi r/q}-e^{-i2\pi p/q}\,z),\qquad z=e^{i\alpha}. \tag{3.45}$$

Thus, $Q_r(z)$ is a polynomial of degree not higher than $q-1$. By (3.45),

$$Q_r(e^{i2\pi(k+r)/q})=e^{-i2\pi r/q}\prod_{s\neq k}(1-e^{-i2\pi s/q})=e^{-i2\pi r/q}q,\qquad 0\le k\le q-1. \tag{3.46}$$

This relation shows that

$$Q_r(z)=e^{-i2\pi r/q}q. \tag{3.47}$$

By (3.38), we have

$$\prod_{p=0}^{q-1}(e^{i2\pi r/q}-e^{-i2\pi p/q}e^{i\alpha})=1-e^{i\alpha q}, \tag{3.48}$$

which in conjunction with (3.44) and (3.47) yields (3.43). □

6.3.3. Now let us consider the case when condition (3.23) is not fulfilled, i.e.,

$$e^{i\alpha_k} = e^{i2\pi r/q_k}, \qquad 0 \le r \le q_k - 1. \tag{3.49}$$

In this case (3.26) remains valid and

$$s_{u,v}^{(k)} = \begin{cases} 0, & \text{for } u+v \not\equiv r \pmod{q_k} \\ q_k e^{-i2\pi u/q_k}, & \text{for } u+v \not\equiv r \pmod{q_k}. \end{cases}$$

Let us introduce the matrix

$$S_k = \left\{ s_{v,u}^{(k)} \right\}_{v,u=0}^{q_k-1}. \tag{3.50}$$

It is readily seen that the elements $t_{v,u}$ of the matrix $S_k^{-1} = T_k$ are given by

$$t_{v,u}^{(k)} = \begin{cases} 0, & \text{for } u+v \not\equiv r \pmod{q_k} \\ \dfrac{e^{-i2\pi u/q_k}}{q_k}, & \text{for } u+v \not\equiv r \pmod{q_k}. \end{cases} \tag{3.51}$$

Denote by e the vector

$$e = \operatorname{col}[1, \cdots, 1].$$

Then, by (3.51),

$$Te = e^{i2\pi r/q_k} \cdot \operatorname{col}\left[1, e^{-i2\pi/q_k}, \cdots, e^{-i2\pi(q_k-1)/q_k}\right]. \tag{3.52}$$

6.3.4. In what follows an essential role will be played by the matrix

$$\widehat{B}(z) = \widehat{U}^*(z) J \widehat{U}(z) \tag{3.53}$$

With this notation, we can rewrite (3.19) in the form

$$v(z) = i\widehat{U}(z)\widehat{j}\widehat{B}^{-1}(z)\widehat{U}^*(z), \qquad \operatorname{Im} z > 0. \tag{3.54}$$

Let us write the matrix $\widehat{B}(z)$ in the block form

$$\widehat{B}(z) = \{b_{p,r}(z)\}_{p,r=1}^{N+1}, \tag{3.55}$$

where $b_{r,r}(z)$ are $q_r \times q_r$ matrices for $1 \le r \le N$, and $b_{N+1,N+1}(z)$ is a $2Q_2 \times 2Q_2$ matrix.

Lemma 3.3. *The matrix-valued function $\widehat{B}(z)$ defined by (3.53) satisfies the following relations:*

$$b_{p,r}(z) = O(z - z_0), \qquad p \ne r, \qquad 1 \le p, r \le N+1, \tag{3.56}$$

$$b_{k,k}(z) = (z - z_0)^{(q_k-1)/q_k} c_k S_k + O(|z - z_0|), \qquad 1 \le k \le N, \tag{3.57}$$

where the matrices S_k, $1 \le k \le N$, are defined by (3.27), (3.28), and the coefficients c_k are defined by (3.5). Moreover,

$$\det S_k \ne 0, \qquad 1 \le k \le N. \tag{3.58}$$

Proof. Relations (3.56) are consequences of Theorems 1.2 and 1.3, while (3.57) follows from (3.26)–(3.28). Finally, (3.58) follows from Lemma 3.1 and formula (3.51). The lemma is proved. □

Now let us examine the block $b_{N+1,N+1}(z)$ of the matrix $\widehat{B}(z)$. To this aim we introduce the matrix

$$S_{N+1} = \begin{bmatrix} M_+ & M \\ M^* & M_- \end{bmatrix}, \tag{3.59}$$

whose blocks, according to (3.17), (3.18), have the form

$$M_\pm = \left\{ \left(Jh^\pm_{Q_1+p}(z_0), h^\pm_{Q_1+r}(z_0) \right) \right\}_{p,r=1}^{2m-Q_1}, \tag{3.60}$$

$$M = \left\{ \left(Jh^+_{Q_1+p}(z_0), h^-_{Q_1+r}(z_0) \right) \right\}_{p,r=1}^{2m-Q_1}. \tag{3.61}$$

It follows from relation (2.9) that

$$(J\xi, \xi) = 0, \qquad \xi \in L^\pm(z_0), \qquad z_0 = \bar{z}_0. \tag{3.62}$$

Finally, by (3.60)–(3.62), one has that $M_+ = M_- = 0$, i.e., the matrix S_{N+1} has the form

$$S_{N+1} = \begin{bmatrix} 0 & M \\ M^* & 0 \end{bmatrix}. \tag{3.63}$$

Lemma 3.4. *The following relations hold:*

$$b_{N+1,N+1}(z) = S_{N+1} + O(z - z_0), \tag{3.64}$$

$$\det S_{N+1} \neq 0. \tag{3.65}$$

Proof. Relation (3.64) follows from the fact that the vector-valued functions $h^\pm_{Q_1+k}(z)$ are analytic. The maximal dimension of an isotropic (i.e., J-orthogonal to itself) subspace in the space $L = L^+(z_0) + L^-(z_0)$ is equal to $2m - Q_1$. Hence, by (3.63), $\det M \neq 0$, which yields (3.65) and completes the proof of the lemma. □

By Lemmas 3.3 and 3.4, the matrix $\widehat{B}(z)$ can be written in the form

$$\widehat{B}(z) = B_0(z)[I + o(1)] = [I + o(1)]B_0(z), \tag{3.66}$$

where

$$B_0(z) = \operatorname{diag}\{S_1 c_1 (z - z_0)^{(q_1-1)/q_1}, \cdots, S_N c_N (z - z_0)^{(q_N-1)/q_N}, S_{N+1}\}. \tag{3.67}$$

From (3.66) we obtain the formula

$$\widehat{B}^{-1}(z) = B_0^{-1}(z)[I + o(1)] = [I + o(1)]B_0^{-1}(z). \tag{3.68}$$

Proof of Theorem 3.1. Relation (3.21) is an immediate consequence of relations (3.54) and (3.66), (3.67), the matrix A (3.21) having the form

$$A = i \sum_{k=1}^{N_1} \frac{U_k(z_0) j_k S_k^{-1} U_k^*(z_0)}{c_k q_1}. \tag{3.69}$$

Here matrices $U_k(z_0)$ in (3.69) are defined by relation (3.20) and

$$j_k = \text{diag } \{\varepsilon_{1,k}, \varepsilon_{2,k}, \cdots, \varepsilon_{q_1,k}\}, \qquad 1 \leq k \leq N_1.$$

In the cases (3.23) and (3.49), the matrices S_k are given by relations (3.27), (3.28) and (3.50), respectively, and the numbers c_k are given by relation (3.5). It follows from (1.19) that

$$(Jh_{k,s}, h_{k,p}) = e^{i\alpha_k p} c_k, \qquad p + s = q_k - 1. \tag{3.70}$$

Since $(Jh_{k,s}, h_{k,p}) = \overline{(Jh_{k,p}, h_{k,s})}$, relations (1.10) and (3.70) yield the equality

$$e^{i\alpha_k(q_k - 1)} = \bar{c}_k / c_k. \tag{3.71}$$

Further, from (3.43) and (3.71) it follows that the formula

$$U_k(z_0) j_k S_k^{-1} U_k^*(z_0) = (\eta_{k,0}, \eta_{k,0}^*) \sum_{s=0}^{q_k} e^{i2\pi(s-1)/q_k} \varepsilon_{r,k} / c_k q_k \tag{3.72}$$

holds in the case (3.23). Thanks to (3.52) and (3.71), formula (3.72) remains valid for the case (3.49) as well. Using formula (1.2) and the connection between $\varepsilon_{r,k}$ and the kind of the corresponding eigenvalue $\rho_k(z)$, we obtain the double inequality

$$-\text{arg } \rho_{k,0} - \frac{\pi}{2} \leq \text{arg } \left(c_{k,1} e^{i(r-1)2\pi/q_k} \varepsilon_{r,k}\right) \leq -\text{arg } \rho_{k,0} + \frac{\pi}{2}. \tag{3.73}$$

Since $c_{k,1} \neq 0$, (3.73) guarantees that

$$\sum_{s=0}^{q_k} e^{-i2\pi(s-1)/q_k} \varepsilon_{r,k} \neq 0. \tag{3.74}$$

Finally, since the vectors $\eta_{k,0}$ $(1 \leq k \leq N_1)$ are linearly independent, relation (3.22) is seen to be a consequence of (3.69), (3.72) and (3.74). The theorem is proved. □

Chapter 7
Division of Canonical Systems into Subclasses

Chapter 4 of this book dealt with the following inverse spectral problem: Given spectral data $\tau(u)$ and α of a canonical system,

$$\frac{d\mathcal{W}(x,z)}{dx} = izJ\mathcal{H}(x)\mathcal{W}(x,z), \tag{0.1}$$

find the Hamiltonian $\mathcal{H}(x)$. However, this formulation of the inverse spectral problem is too general. We must divide the class of canonical systems into subclasses, and then together with the spectral data $\tau(u)$ and α we must specify the subclass in which we are looking for the corresponding Hamiltonian. This chapter is devoted to the realization of this program.

7.1. An effective solution of the inverse problem

7.1.1. In this subsection we indicate conditions under which the process of solving the inverse problem of Chapter 4.2 reduces to the solution of a system of linear algebraic equations. We note that a similar result for the classical inverse problem [56] plays an important role in the theory of solitons [62].

Consider a symmetric S-node

$$AS - SA^* = i(\Phi_1\Phi_2^* + \Phi_2\Phi_1^*), \tag{1.1}$$

where $S, A \in \{\mathfrak{H}, \mathfrak{H}\}$, $\Phi_1, \Phi_2 \in \{\mathfrak{G}_1, \mathfrak{H}\}$. We shall assume that the S-node (1.1) satisfies the following conditions:

I. The operator S is positive and invertible.

II. There exists a continuous increasing family of orthogonal projectors P_ζ, $(0 \le \zeta \le l$, $P_0 = 0$, $P_l = I)$ such that $A^*P_\zeta = P_\zeta A^* P_\zeta$, $0 \le \zeta \le l$.

III. The spectrum of the operator A is concentrated at the origin and there exists a constant $M > 0$ such that

$$\|(P_{\zeta+\Delta\zeta} - P_\zeta)A(P_{\zeta+\Delta\zeta} - P_\zeta)\| \le M|\Delta\zeta|, \qquad 0 \le \zeta \le l.$$

IV. Zero is not an eigenvalue of A.

V. The ranges of A and Φ_2 have null intersection.

VI. $\Phi_2 g = 0$ implies $g = 0$.

By Theorem 2.6 of Chapter 4, to the operator identity (1.1) one can associate the system

$$Y(x,z) = Y(0,z) + izJ\int_0^x [dB_1(t)]Y(t,z), \tag{1.2}$$

where

$$Y(x,z) = \mathrm{col}\,[Y_1(x,z), Y_2(x,z)], \qquad Y_1(0,z) = 0.$$

We will assume in addition that the following condition is satisfied:

VII. The matrix-valued function $B_1(t)$ is absolutely continuous.

Setting

$$\mathcal{H}(x) = B_1'(x) \tag{1.3}$$

we pass from the system (1.2) to the canonical system (0.1).

Let $\tau(u)$ be a spectral matrix function of the system (0.1). Consider a monotonically increasing $m \times m$ matrix-valued function $\widetilde{\tau}(u)$ $(-\infty < u < \infty)$ such that

$$\widetilde{\tau}(u) = \tau(u) + \gamma(u), \tag{1.4}$$

where $\gamma(u)$ is a piecewise constant matrix-valued function with jumps γ_j at points u_j $(1 \le j \le m < \infty)$. Then the following operator is correctly defined:

$$\widetilde{S} = \int_{-\infty}^{\infty} (I - Au)^{-1}\Phi_2[d\widetilde{\tau}(u)]\Phi_2^*(I - A^*u)^{-1}. \tag{1.5}$$

Formula (1.5) can also be written in the form

$$\widetilde{S} = S + \sum_{j=1}^{m}(I - Au_j)^{-1}\Phi_2\gamma_j\Phi_2^*(I - A^*u_j)^{-1}. \tag{1.6}$$

It follows from (1.6) that

$$\operatorname{rank}(\widetilde{S} - S) \le p \cdot \dim \mathfrak{G}_1 < \infty.$$

We shall say that the matrix-valued function $\widetilde{\tau}(u)$ is *p-regular* if conditions I–VII and (1.4) are satisfied and $\widetilde{S}$ is an invertible operator. Applying Theorem 2.6 of Chapter 4, we obtain the following assertion.

Proposition 1.1. *If the matrix-valued function $\widetilde{\tau}(u)$ is p-regular, then it is a spectral matrix function for a system*

$$Y(x, z) = Y(0, z) + izJ\int_0^x [d\widetilde{B}_1(t)]Y(t, z), \tag{1.7}$$

with the boundary condition

$$Y_1(0, z) = 0. \tag{1.8}$$

Moreover, relations similar to relations (1.13) *of Chapter* 4 *hold, and also*

$$\widetilde{B}_1(\xi) = \widetilde{\Pi}^*\widetilde{S}_\xi^{-1}P_\xi\widetilde{\Pi}, \qquad \widetilde{\Pi} = [\widetilde{\Phi}_1 \ \ \Phi_2], \qquad \widetilde{S}_\xi = P_\xi\widetilde{S}P_\xi, \tag{1.9}$$

$$\widetilde{\Phi}_1 = \Phi_1 - i\sum_{j=1}^{p}\left[A(I - u_jA)^{-1} + \frac{u_j}{1 + u_j^2}I\right]\Phi_2\gamma_j + i\Phi_2\alpha, \qquad \alpha = \alpha^*. \tag{1.10}$$

Remark 1.1. If S_ξ^{-1} $(0 < \xi \leq l)$ is known, then, according to (1.5) and (1.6), the construction of $\widetilde{S}_\xi^{-1}$ reduces to the solution of a system of linear equations. Hence, thanks to relations (1.9) and (1.10), the problem of recovering $\widetilde{B}_1(\xi)$ from $\widetilde{\tau}(u)$ can be solved explicitly.

7.1.2. Let us examine separately the case $p = 1$. In this case $\gamma(u) = \widetilde{\tau}(u) - \tau(u)$ has a jump γ at only one point $u_1 = \lambda$.

Denote

$$L(\xi, \lambda) = I + \Phi_2^*(I - A^*\lambda)^{-1}S_\xi^{-1}(I - Al)^{-1}P_\xi\Phi_2\gamma. \tag{1.11}$$

Proposition 1.2. *Let $\widetilde{\tau}(u)$ be a 1-regular matrix-valued function. If for any ξ, $0 < \xi \leq l$, the equation $P_\xi\Phi_2 g = 0$ has only the trivial solution, then*

$$\det L(\xi, \lambda) \geq 1 \tag{1.12}$$

and

$$\widetilde{S}_\xi^{-1} = S_\xi^{-1} - S_\xi^{-1}(I - A\lambda)^{-1}\Phi_2\gamma L^{-1}(\xi, \lambda)\Phi_2^*(I - A^*\lambda)^{-1}S_\xi^{-1}. \tag{1.13}$$

Proof. Suppose first that $\gamma > 0$. Then from (1.11) we obtain

$$\gamma^{\frac{1}{2}}L(\xi, \lambda)\gamma^{-\frac{1}{2}} = I + B, \qquad B \geq 0,$$

i.e., (1.12) holds. Taking $\gamma_\varepsilon = \gamma + \varepsilon I_m$ and letting $\varepsilon \to 0$, we conclude that (1.12) remains valid when $\gamma \geq 0$.

By (1.6), we have

$$\widetilde{S}_\xi = S_\xi + P_\xi(I - A\lambda)^{-1}\Phi_2\gamma\Phi_2^*(I - A^*\lambda)^{-1}P_\xi. \tag{1.14}$$

Multiplying the right sides of (1.13) and (1.14) and recalling (1.11) we see that (1.13) holds, which completes the proof of the proposition. □

Next, let us clarify how the matrix-valued functions

$$\mathcal{W}(\xi, z) = I + izJ\Pi^*S_\xi^{-1}(I - zA)^{-1}P_\xi\Pi \tag{1.15}$$

and

$$\widetilde{\mathcal{W}}(\xi, z) = I + izJ\widetilde{\Pi}^*\widetilde{S}_\xi^{-1}(I - zA)^{-1}P_\xi\widetilde{\Pi} \tag{1.16}$$

are related. By analogy with relation (3.1) of Chapter 2, we obtain from the (1.15) the equality

$$\mathcal{W}^*(\xi, z)J\mathcal{W}(\xi, \zeta) - J = i(\zeta - \bar{z})\Pi^*(I - zA^*)^{-1}S_\xi^{-1}(I - \xi A)^{-1}P_\xi\Pi. \tag{1.17}$$

Consider the block representation

$$\mathcal{W}^*(\xi, \bar{z}) = \begin{bmatrix} a(\xi, z) & b(\xi, z) \\ c(\xi, z) & d(\xi, z) \end{bmatrix}. \tag{1.18}$$

Further, introduce the matrix-valued functions

$$U(x, z, \lambda) = I + \frac{z}{z-\lambda} \cdot \frac{i}{\lambda} Q(x, \lambda), \qquad \lambda = \bar{\lambda} \neq 0, \tag{1.19}$$

$$Q(x, \lambda) = \begin{bmatrix} c^*(x, \lambda) \\ d^*(x, \lambda) \end{bmatrix} \gamma L^{-1}(x, \lambda)[d(x, \lambda) \;\; c(x, \lambda)], \tag{1.20}$$

$$V(\lambda) = \begin{bmatrix} I_m & 0 \\ -i\gamma/l & I_m \end{bmatrix}. \tag{1.21}$$

Proposition 1.3. *The matrix-valued functions $V(z)$ and $U(x, z, \lambda)$ are J-unitary for all z such that $z = \bar{z}$, i.e.,*

$$V^*(z)JV(z) = J, \qquad U^*(x, z, \lambda)JU(x, z, \lambda) = J. \tag{1.22}$$

Proof. The first of relations (1.22) can be verified directly. To prove the second of relations (1.22), we observe that

$$c(x, \lambda)d^*(x, \lambda) + d(x, \lambda)c^*(x, \lambda) = 0, \tag{1.23}$$

which follows from (1.17) and (1.18) upon setting $\zeta = z = \lambda$ and $\xi = x$. Also, by (1.11),

$$\gamma L^{-1}(x, \lambda) = L^{*-1}(x, \lambda)\gamma. \tag{1.24}$$

The needed relation is seen to be a consequence of (1.19), (1.23), and (1.24). □

We are now ready to prove the following assertion.

Theorem 1.1. *Let $\widetilde{\tau}(u)$ be a 1-regular matrix-valued function and let $\alpha = \frac{\lambda\gamma}{1+\lambda^2}$, with $\lambda \neq 0$ (see (1.10)). If for any ξ, $0 < \xi \le l$, the equation $P_\xi \Phi_2 g = 0$ has only the trivial solution, then*

$$\widetilde{\mathcal{W}}(x, z) = V(\lambda)U(x, z, \lambda)V(z - \lambda). \tag{1.25}$$

Proof. By (1.10),

$$\widetilde{\Phi}_1 = \Phi_1 - iA(I - lA)^{-1}\Phi_2\gamma. \tag{1.26}$$

Using formulas (1.13) and (1.16), we write

$$\widetilde{\mathcal{W}}(\xi, z) = I + izJ\widetilde{\Pi}^*[S_\xi^{-1} - S_\xi^{-1}(I - Al)^{-1}\Phi_2\gamma L^{-1}(\xi, \lambda)\Phi_2^*(I - A^*\lambda)^{-1}S_\xi^{-1}] \times$$
$$\times (I - Az)^{-1}P_\xi\widetilde{\Pi}. \tag{1.27}$$

Relation (1.25) follows directly from (1.27) if one recalls (1.15), (1.17), (1.18) and the equality

$$\widetilde{\Pi} = \begin{bmatrix} \Phi_1 - iA(I - \lambda A)^{-1}\Phi_2\gamma \\ \Phi_2 \end{bmatrix}.$$

This proves the theorem. □

By Theorem 1.2 of Chapter 3 on continual factorization, we have

$$\mathcal{W}(\xi, z) = \int_0^{\xi} \exp[izJdB(t)], \qquad \widetilde{\mathcal{W}}(\xi, z) = \int_0^{\xi} \exp[izJd\widetilde{B}(t)],$$

where

$$B(\xi) = \Pi^* S_\xi^{-1} P_\xi \Pi, \qquad \widetilde{B}(\xi) = \widetilde{\Pi}^* \widetilde{S}_\xi^{-1} P_\xi \widetilde{\Pi}.$$

Proposition 1.4. *Assume that the conditions of Proposition* 1.2 *are satisfied and that* $B(x)$ *is absolutely continuous on* $[0, l]$. *Then for* $\lambda \neq 0$ *the matrix-valued functions* $\mathcal{W}(x, z)$, $U(x, z, l)$, $\widetilde{\mathcal{W}}(x, z)$, *and* $\widetilde{B}(x)$ *are also absolutely continuous on* $[0, l]$.

Proof. It follows from (1.17) that

$$\Phi_2^*(I - A^*\lambda)^{-1} S_\xi^{-1} (I - A\lambda)^{-1} P_\xi \Phi_2 = -\left[d(\xi, \lambda) \frac{\partial}{\partial \lambda} c^*(\xi, \lambda) + c(\xi, \lambda) \frac{\partial}{\partial \lambda} d^*(\xi, \lambda)\right]. \tag{1.28}$$

Now the fact that $\mathcal{W}(x, z)$ and $\dfrac{\partial}{\partial z}\mathcal{W}(x, z)$ are absolutely continuous follows from the equality

$$\frac{d\mathcal{W}(x, z)}{dx} = izJ\mathcal{H}(x)\mathcal{W}(x, z), \tag{1.29}$$

where $\mathcal{H}(x) = B'(x)$. Further, from (1.11) and (1.28) we conclude that $L(x, \lambda)$ and $L^{-1}(x, \lambda)$ are absolutely continuous. By (1.19) and (1.20), the matrix-valued function $U(x, z, \lambda)$ is absolutely continuous. Using (1.25), we see that $\widetilde{\mathcal{W}}(x, z)$ is absolutely continuous, and so is $\widetilde{B}(x)$. □

7.1.3. Let us introduce the notation

$$\widetilde{U}(x) = V(\lambda) \cdot \lim_{z \to \infty} U(x, z, \lambda) = V(\lambda)\left[I + \frac{i}{\lambda} Q(x, \lambda)\right]. \tag{1.30}$$

It follows from Proposition 1.3 that the matrix-valued function $\widetilde{U}(x)$ is J-unitary. Consequently, $\widetilde{U}(x)$ is invertible for every $x \in [0, l]$.

Theorem 1.2. *Let the following conditions be satisfied:*

1) *The matrix-valued functions* $\widetilde{\tau}(u)$ *is* 1*-regular.*
2) *For any* ξ, $0 < \xi \leq l$, *the equation* $P_\xi \Phi_2 g = 0$ *has only the trivial solution.*
3) $\lambda \neq 0$ *and* $B(x)$ *is absolutely continuous on* $[0, l]$.

Then for almost every $x \in [0, l]$

$$J\widetilde{B}'(x) = \widetilde{U}(x) J B'(x) \widetilde{U}^{-1}(x). \tag{1.31}$$

Proof. By relations (1.16) and (1.14) of Chapter 4, we have

$$\frac{d\widetilde{\mathcal{W}}(x,z)}{dx} = izJ\widetilde{B}'(x)V(\lambda)U(x,z,\lambda)\mathcal{W}(x,z)V(z-\lambda).$$

On the other hand, from (1.25) it follows that

$$\frac{d\widetilde{\mathcal{W}}(x,z)}{dx} = V(\lambda)\frac{d}{dx}U(x,z,\lambda)\mathcal{W}(x,z)V(z-\lambda)+$$

$$+V(\lambda)U(x,z,l)izJB'(x)\mathcal{W}(x,z)V(z-\lambda). \tag{1.32}$$

Thus

$$izJ\widetilde{B}'(x)V(\lambda)U(x,z,\lambda) = V(\lambda)\frac{d}{dx}U(x,z,\lambda)+V(\lambda)U(x,z,\lambda)izJB'(x). \tag{1.33}$$

Dividing both sides of (1.26) by z and letting $z \to \infty$, we obtain (1.31). □

Proposition 1.5. *Assume that the conditions 2), 3) of Theorem 1.2 are satisfied and that $\widetilde{\tau}(u)$ is a p-regular matrix-valued function ($p \geq 1$). Then the perturbed Hamiltonian $J\widetilde{B}'(x)$ is linearly similar to the initial Hamiltonian $JB'(x)$ at almost every point x.*

Remark 1.1. In a number of cases the method of constructing the perturbed Hamiltonian $J\widetilde{\mathcal{H}}(x) = J\widetilde{B}'(x)$ with the help of (1.31) leads to the result faster then the general formulas (1.9),(1.10).

7.2. Two principles of dividing a class of canonical systems into subclasses

7.2.1. We shall say that the Hamiltonian $\mathcal{H}(x)$ of system (0.1) belongs to the *subclass* $\mathfrak{N}(M)$, if the following representation holds with a constant real matrix M:

$$J\mathcal{H}(x) = U(x)MU^{-1}(x). \tag{2.1}$$

Remark 2.1. Case (2.1) is only one of the cases which can be investigated by our method. However, it is a very important case, since it covers the classical problems (the Sturm-Liouville equation, the Dirac-type systems, the string equation).

7.2.2. Let us also introduce the subclasses $\widetilde{\mathfrak{N}}(A,\Phi_2)$ of the class of canonical systems. To this end we consider the operator identities

$$AS - SA^* = i(\Phi_1\Phi_2^* + \Phi_2\Phi_1^*). \tag{2.2}$$

We will assume that the following conditions are satisfied:

I) The operators A and S, with $S > 0$, act in the space $L^2_m(0,\omega)$ by the rules

$$Sf = f(x) + \int_0^\omega K(x,t)f(t)\,dt, \qquad f(\cdot) \in L^2_m(0,\omega), \tag{2.3}$$

and

$$Af = \int_0^x a(x,t)f(t)\,dt, \qquad f(\cdot) \in L^2_m(0,\omega), \tag{2.4}$$

where $K(x,t)$ and $a(x,t)$ are continuous $m \times m$ matrix-valued functions.

II) The operators Φ_1 and Φ_2 have the form

$$\Phi_p g = L_p(x)g, \qquad p = 1, 2, \tag{2.5}$$

where $L_1(x)$ and $L_2(x)$ are continuous $m \times m$ matrix-valued functions and g is an arbitrary constant $m \times 1$ column vector.

Consider the following family of projectors in $L^2_m(0,\omega)$:

$$P_\zeta f = \begin{cases} f, & \text{for } 0 \leq x \leq \zeta \\ 0 & \text{for } 0 < x \leq \omega. \end{cases} \tag{2.6}$$

Then, by definition (2.4),

$$P_\zeta A^* P_\zeta = A^* P_\zeta. \tag{2.7}$$

With identity (2.2) we associate the system (see Chapters 3 and 4)

$$\mathcal{W}(x,z) = I_{2m} + izJ\int_0^x [dB(t)]\mathcal{W}(t,z), \tag{2.8}$$

where

$$B(\zeta) = \Pi^* S_\zeta^{-1} P_\zeta \Pi, \qquad \Pi = [\Phi_1 \ \ \Phi_2], \qquad S_\zeta = P_\zeta S P_\zeta. \tag{2.9}$$

7.2.3. It follows from condition I) above that the inverse operator S_ζ^{-1} has the form

$$S_\zeta^{-1} f(x) = f(x) + \int_0^\zeta \Gamma_\zeta(x,t)f(t)\,dt, \qquad f(\cdot) \in L^2_m(0,\omega). \tag{2.10}$$

Moreover, by the same condition, the operator S admits a factorization, i.e., can be represented in the form (see [34], Chapter 4)

$$S^{-1} = (I + V_-^*)(I + V_-), \tag{2.11}$$

where

$$V_- f = \int_0^x \Gamma_x(x,t)f(t)\,dt. \tag{2.12}$$

From relations (2.9) and (2.11) we deduce that

$$B'(x) = \mathcal{H}(x) = \{F_p^*(x)F_q(x)\}_{q,p=1}^2, \tag{2.13}$$

where

$$F_p(x) = (I + V_-)L_p(x), \quad p = 1, 2. \tag{2.14}$$

Here the operator $I + V_-$ acts on the matrix-valued function $L_p(\cdot)$ columnwise. Note that the matrix-valued function $\Gamma_x(x,t)$ in (2.12) is continuous ([34], Ch. 4). Hence, by (2.13) and (2.14), the matrix-valued function $\mathcal{H}(x)$ is also continuous. It follows from the absolute continuity of $B(x)$ that the system (2.8) is equivalent to the system (0.1).

7.2.4. We can now introduce the subclasses $\widetilde{\mathbf{N}}(A, \Phi_2)$. Fix the operators A and Φ_2. Denote by $\widetilde{\mathbf{N}}(A, \Phi_2)$ the class of Hamiltonians $\mathcal{H}(x)$ which is obtained when one allows Φ_1 and S to vary. Here one assumes that relations (2.3) and (2.5) hold.

Conjecture 2.1. *If $\mathcal{H}_1(x)$ and $\mathcal{H}_2(x)$ belong to $\widetilde{\mathbf{N}}(A, \Phi_2)$, then the matrix-valued functions $J\mathcal{H}_1(x)$ and $J\mathcal{H}_2(x)$ are linearly similar for every x.*

This conjecture turns out to be true under the additional strong requirement introduced in Section 7.1. In this section we prove Conjecture 2.1 for a number of classes $\widetilde{\mathbf{N}}(A, \Phi_2)$ which are important in different applications. To this end we use the results of Section 7.1.

Theorem 2.1. *Let the following conditions be satisfied:*

1) *The operators S, A, and Φ_k are defined by formulas (2.3)–(2.5) and S_0, Φ_1^0 have the form*

$$S_0 f = f(x) + \int_0^\omega K_0(x,t) f(t)\, dt, \tag{2.15}$$

$$\Phi_1^0 g = L_1^0(x) g, \tag{2.16}$$

where $K_0(x,t)$ and $L_1^0(x)$ are continuous $m \times m$ matrix-valued functions.

2) *The matrix $J\mathcal{H}_0(x)$ defined by relations (2.13)–(2.16) is linearly similar to a diagonal matrix $D(x)$.*

3) *There exist real numbers $u_j(r)$ and selfadjoint matrices $\gamma_j(r)$ and $\alpha_{N(r)}$ such that the following inequalities hold for $\varepsilon_r \to 0$:*

$$\sup \|K(x,t) - K(x,t,r)\| < \varepsilon_r, \qquad 0 \le x, t \le \omega, \tag{2.17}$$

$$\sup \|L_1(x) - L_1(x,r)\| < \varepsilon_r, \qquad 0 \le x, t \le \omega. \tag{2.18}$$

Here

$$L_1(x,r) = L_1^0(x) - \sum_{j=1}^{N(r)} A(I - u_j(r)A)^{-1} L_2(x)\gamma_j(r) - iL_2(x)\alpha_{N(r)} \tag{2.19}$$

and $K(x,t,r)$ is defined by the equation

$$\begin{aligned} S(r)f &= f + \int_0^\omega K(x,t,r)f(t)dt \\ &= S_0 f + \sum_{j=1}^{N(r)} (I - u_j(r)A)^{-1}\Phi_2\gamma_j(r)\Phi_2^*(I - u_j(r)A^*)^{-1} f. \end{aligned} \tag{2.20}$$

Then the matrix $J\mathcal{H}(x)$ defined by relations (2.3), (2.5), (2.13) *is linearly similar to $J\mathcal{H}_0(x)$. (The norms of the matrices figuring in the left-hand sides of* (2.17) *and* (2.18) *are defined by the metric of the space $\mathfrak{G}_1$.)*

Proof. The sequence of operators $S(r)$ satisfy the identities

$$AS(r) - S(r)A^* = i(\Phi_1(r)\Phi_2^* + \Phi_2\Phi_1^*(r)), \tag{2.21}$$

where

$$\Phi_1(r)g = L_1(x,r)g. \tag{2.22}$$

The system

$$\frac{d\mathcal{W}}{dx} = izJ\mathcal{H}(x)\mathcal{W}$$

belonging to the class $\widetilde{\mathbf{N}}(A,\Phi_2)$ corresponds to identity (2.21). By analogy to (2.13) and (2.14), we deduce the equalities

$$\mathcal{H}(x,r) = \{F_p^*(x,r)F_q(x,r)\}_{q,p=1}^2, \tag{2.23}$$

with

$$F_p(x,r) = (I + V_-(r))L_p(x,r), \tag{2.24}$$

and

$$V_-(r)f = \int_0^x \Gamma_x(x,t,r)f(t)\,dt. \tag{2.25}$$

It follows from the estimate (2.17) (see [34], Ch. 4) that

$$\Gamma_x(x,t,r) \to \Gamma_x(x,t), \quad \varepsilon_r \to 0, \tag{2.26}$$

which in conjunction with (2.18) and (2.24) shows that

$$F_p(x,r) \to F_p(x), \qquad p = 1,2, \qquad \varepsilon_r \to 0. \tag{2.27}$$

Therefore, by (2.13) and (2.23), we have

$$\mathcal{H}(x,r) \to \mathcal{H}(x), \qquad \varepsilon_r \to 0. \tag{2.28}$$

The matrices $J\mathcal{H}(x,r)$ and $J\mathcal{H}_0(x)$ are linearly similar (Theorem 1.2). Hence, by equality (2.28) and condition 2) of the theorem, the matrices $J\mathcal{H}(x)$ and $J\mathcal{H}_0(x)$ are also linearly similar. The theorem is proved. □

Example 1.2. Let us consider the class $\widetilde{\mathbf{N}}(A, \Phi_2)$, where the operators are chosen in the following way:

$$Af = iW\int_0^x f(t)\,dt, \quad \Phi_2 g = g, \quad W = \operatorname{diag}\{w_1, w_2, \dots, w_m\}, \qquad w_k > 0. \tag{2.29}$$

Then the kernel $K(x,t)$ has the following structure:

$$K(x,t) = \{k_{p,q}(w_p x - w_q t)\}_{p,q=1}^2, \qquad K(x,t) = K^*(t,x),$$

and the matrix $L_1(x)$ is defined by the relation

$$L_1(x) = W\int_0^x K(u,0)\,du + \frac{1}{2}W. \tag{2.30}$$

Setting $S = I$, formula (2.13) yields

$$\mathcal{H}_0(x) = \begin{bmatrix} \frac{1}{4}W^2 & \frac{1}{2}W \\ \frac{1}{2}W & I_m \end{bmatrix}. \tag{2.31}$$

Corollary 2.1. *Let A and Φ_2 be defined by equalities (2.29) and $\mathcal{H}_0(x)$ be defined by formula (2.31). If $\mathcal{H}(x) \in \widetilde{\mathbf{N}}(A, \Phi_2)$, then the matrices $J\mathcal{H}(x)$ and $J\mathcal{H}_0(x)$ are linearly similar.*

Proof. Indeed, condition (2.17) of Theorem 2.1 means that the kernel $K(x,t)$ can be uniformly approximated by linear combinations of kernels of the operators

$$\mathcal{K}_j f = (I - u_j A)^{-1}\Phi_2\gamma_j\Phi_2^*(I - u_j A^*)^{-1} f = \int_0^\omega K_j(x,t) f(t)\,dt\ . \tag{2.32}$$

In the case (2.29), we have

$$K_j(x,t) = e^{iu_jWx}\gamma_j e^{-iu_jWt}.$$

The fact that condition (2.17) is satisfied follows from the theory of trigonometric series [100]. Condition (2.18) is verified in a similar manner. Let us also remark that $J\mathcal{H}_0(x)$ is linearly similar to the diagonal matrix

$$M_1 = \begin{bmatrix} W & 0 \\ 0 & 0 \end{bmatrix}. \tag{2.33}$$

This completes the proof of Corollary 2.1. □

In this section we have introduced two methods for singling out subclasses $\widetilde{\mathfrak{N}}(A,\Phi_2)$ and $\mathfrak{N}(M)$ of the class of canonical systems. If the conditions of Corollary 2.1 are fulfilled, then

$$\widetilde{\mathfrak{N}}(A,\Phi_2) \subset \mathfrak{N}(M_1).$$

Example 2.2. Choose A and Φ_2 as follows:

$$Af = \int_0^x (t-x)f(t)\,dt, \qquad \Phi_2 g = g. \tag{2.34}$$

Then the kernel $K(x,t)$ has the following structure [84]:

$$K(x,t) = R(x-t) + R(x+t), \tag{2.35}$$

where

$$R(x) = R^*(x) = R(-x) \tag{2.36}$$

and $L_1(x)$ is given by

$$L_1(x) = i\left[2\int_0^x (x-u)R(u)\,du + \frac{1}{2}x\right]. \tag{2.37}$$

Setting $S_0 = I$, formula (2.13) yields

$$\mathcal{H}_0(x) = \begin{bmatrix} x^2 I_m & -ixI_m \\ ixI_m & I_m \end{bmatrix}. \tag{2.38}$$

By (2.13) and (2.14), the matrix $\mathcal{H}(x)$ can be written in the form

$$\mathcal{H}(x) = \begin{bmatrix} F_1^*(x) \\ F_2^*(x) \end{bmatrix} [F_1(x)\ \ F_2(x)], \tag{2.39}$$

where

$$F_2(x) = (I+V_-)I_m. \tag{2.40}$$

It follows from (2.39) that

$$\operatorname{rank}\mathcal{H}(x) \le m. \tag{2.41}$$

In Example 2.2 condition 2) of Theorem 2.1 is not satisfied. In this case the matrix $J\mathcal{H}_0(x)$ is linearly similar to the matrix

$$M_2 = \begin{bmatrix} 0 & 0 \\ I_m & 0 \end{bmatrix},$$

i.e., $J\mathcal{H}_0(x)$ cannot be reduced to diagonal form. However the following assertion is true in this case as well.

Corollary 2.2. *Let A and Φ_2 be defined by equalities* (2.34) *and $\mathcal{H}_0(x)$ be defined by equality* (2.38). *If $\mathcal{H}(x) \in \widetilde{\mathbf{N}}(A, \Phi_2)$ and*

$$\operatorname{rank} \mathcal{H}(x) = m, \tag{2.42}$$

then the matrices $J\mathcal{H}(x)$ and $J\mathcal{H}_0(x)$ are linearly similar.

Proof. Indeed, here the kernel $K_j(x,t)$ of the operator $\mathcal{K}_j$ (see (2.32)) has the form

$$K_j(x,t) = \gamma_j \cos(\sqrt{u_j}\, x) \cos(\sqrt{u_j}\, t). \tag{2.43}$$

Using again the theory of the trigonometric series we can verify as above that conditions (2.17) and (2.18) of Theorem 2.1 are satisfied. Hence, (2.28) holds. Since the matrix $J\mathcal{H}(x,r)$ is similar to M_2, we obtain the relation

$$[J\mathcal{H}(x,r)]^2 = 0 \tag{2.44}$$

Therefore, by (2.28) and (2.44),

$$[J\mathcal{H}(x)]^2 = 0, \tag{2.45}$$

which in conjunction with (2.42) shows that the matrix $J\mathcal{H}(x)$ is linearly similar to M_2, as claimed □

As in the previous example,

$$\widetilde{\mathbf{N}}(A, \Phi_2) \subset \mathbf{N}(M_2).$$

Chapter 8
Uniqueness Theorems

In Chapter 4 we described a method of constructing a canonical system from known spectral data. However, different canonical systems may correspond to the same spectral data, i.e., in the general case uniqueness fails to hold.

In the present chapter we single out subclasses of canonical systems for which a uniqueness theorem holds. In Section 1 we obtain uniqueness theorems under the assumption that the monodromy matrix $W(z) = \mathcal{W}(l, z)$ is known. Let us remark that Theorem 1.1 of Chapter 3 gives a method of constructing a canonical system from a known monodromy matrix $W(z)$.

In Section 2 we consider the case in which the spectral data $\tau(u)$ and α are given.

8.1. Monodromy matrix and uniqueness theorems

8.1.1. Let us begin by considering systems of the form

$$\frac{dY}{dx} = [\lambda D + A(x)]Y, \qquad 0 \le x \le l, \tag{1.1}$$

where $A(x)$ and $B(x)$ are $n\times n$ matrices and $Y(x)$ is a $n\times 1$ matrix. The monodromy matrix $W_A(\lambda)$ of system (1.1) is defined by the relation

$$W_A(\lambda)Y(0, \lambda) = Y(l, \lambda). \tag{1.2}$$

We shall assume that the matrix D is diagonal and has the form

$$D = \text{diag}\,\{D_1, D_2, \cdots, D_r\}, \tag{1.3}$$

where

$$D_j = \alpha_j I_{m_j}, \qquad \sum_{j=1}^{r} m_j = n, \qquad \alpha_j \ne \alpha_k, \quad j \ne k.$$

Together with system (1.1) we shall consider the system

$$\frac{dY}{dx} = [\lambda D + B(x)]Y, \qquad 0 \le x \le l, \tag{1.4}$$

and its monodromy matrix $W_B(\lambda)$.

Let us decompose the matrices $A(x)$ and $B(x)$ into blocks

$$A(x) = \{A_{ij}(x)\}_{i,j=1}^{r}, \qquad B(x) = \{B_{ij}(x)\}_{i,j=1}^{r},$$

with respect to the decomposition defined by (1.3).

Theorem 1.1 (Uniqueness theorem [91]). *Let the following conditions be satisfied:*

1) *The entries of the matrix-valued functions $A(x)$ and $B(x)$ belong to $L^2(0,l)$.*
2)
$$A_{ii}(x) = B_{ii}(x) = 0, \qquad 1 \leq i \leq r. \tag{1.5}$$
3) *The matrix $W_B(\lambda)W_A^{-1}(\lambda)$ is block-diagonal.*

Then
$$A(x) = B(x). \tag{1.6}$$

Proof. We shall use the system
$$\frac{dQ}{dx} = -[\bar{\lambda}D + B^*(x)]Q, \tag{1.7}$$
where $Q(x,\lambda)$ is an $n \times 1$ matrix-valued function. It is readily verified that
$$Q^*(l,\lambda) = Q^*(0,\lambda)W_B^{-1}(\lambda). \tag{1.8}$$
Set
$$R(x,\lambda) = Y(x,\lambda)Q^*(x,\lambda). \tag{1.9}$$
From (1.1) and (1.7) we obtain the equality
$$\frac{dR}{dx} = \lambda(DR - RD) + A(x)R - RB(x). \tag{1.10}$$
Taking into account (1.2) and (1.8), we deduce from (1.1) and (1.7) the relation
$$\int_0^l Q^*(x,\lambda)[A(x)-B(x)]Y(x,\lambda)\,dx = Q^*(l,\lambda)[I_n - W_B(\lambda)W_A^{-1}(\lambda)]Y(l,\lambda). \tag{1.11}$$
Denote by $\mathfrak{L}$ the space of $n \times n$ matrix-valued functions with the inner product
$$(R_1,R_2)_{\mathfrak{L}} = \int_0^l \mathrm{Tr}\ [R_1(x)R_2^*(x)]\,dx. \tag{1.12}$$
Since $\mathrm{Tr}\ [R_1R_2^*] = \mathrm{Tr}\ [R_2R_1^*]$, it follows from (1.9) and (1.11) that
$$(R,C)_{\mathfrak{L}} = \mathrm{Tr}\ \left\{[I_n - W_B(l)W_A^{-1}(l)]R(l,l)\right\}, \tag{1.13}$$
where
$$C(x) = A^*(x) - B^*(x). \tag{1.14}$$
Next, denote by $\mathfrak{L}_1$ the subspace of block-diagonal matrix-valued functions, with the dimensions of the blocks matching the decomposition (1.3). The subspace $\mathfrak{L}_2 = \mathfrak{L} \ominus \mathfrak{L}_1$ consists of the matrix-valued functions whose diagonal blocks are equal to zero.

Let P_k be the orthogonal projector from $\mathfrak{L}$ onto $\mathfrak{L}_k$. Setting $R_k = P_k R$ we rewrite the system (1.10) in the form

$$\begin{cases} \dfrac{dR_1}{dx} = P_1[A(x)R_2 - R_2B(x)], \\ \dfrac{dR_2}{dx} = \lambda K R_2 + [A(x)R_1 - R_1B(x)] + P_2[A(x)R_2 - R_2B(x)], \end{cases} \tag{1.15}$$

where the operator K acts in $\mathfrak{L}_2$ by the rule

$$KR_2 = DR_2 - R_2D. \tag{1.16}$$

Assume further that

$$R_1(l) = 0. \tag{1.17}$$

Then from (1.15) we derive the following equality:

$$R_1 = T_1R_2 = -\int_x^l P_1[A(u)R_2(u) - R_2(u)B(u)]\,du. \tag{1.18}$$

It follows from (1.13), (1.17) and condition 3) of the theorem that

$$(R_2(x,\lambda), C(x))_{\mathfrak{L}} = 0. \tag{1.19}$$

Differentiating both sides of (1.19) with respect to λ we obtain

$$\left(\frac{\partial^p R_2(x,\lambda)}{\partial \lambda^p}, C(x)\right)_{\mathfrak{L}} = 0, \qquad p = 0, 1, 2, \dots . \tag{1.20}$$

Let us associate with the system (1.15) the operator pencil

$$L(\lambda)Y = Y - \lambda SKY - TY, \qquad Y \in \mathfrak{L}_2, \tag{1.21}$$

where the operators S and T are defined by the relations

$$SR_2 = \frac{1}{2}\left[\int_0^x R_2(u)\,du - \int_x^l R_2(u)\,du\right], \tag{1.22}$$

and

$$TR_2 = S\left\{A(x)(T_1R_2) - (T_1R_2)B(x) + P_2[A(x)R_2 - R_2B(x)]\right\}, \tag{1.23}$$

respectively. Now let $Y_0, Y_1, \dots, Y_m$ be a system of eigenvectors and generalized eigenvectors of the pencil $L(\lambda)$ at the point λ_j. If we set $Y_{-1} = 0$, then we have the relations

$$L(l_j)Y_p = SKY_{p-1}, \qquad p = 0, 1, \dots, m. \tag{1.24}$$

Choose $R_2(\lambda)$ so that

$$\left.\frac{\partial^p R_2(l,\lambda)}{\partial\lambda^p}\right|_{\lambda=\lambda_j} = Y_p(l), \qquad p = 0,1,\dots,m. \tag{1.25}$$

Differentiating both sides of (1.15) with respect to λ and both sides of (1.24) with respect to x we see that $\partial^p R_2(x,\lambda)/\partial\lambda^p\big|_{\lambda=\lambda_j}$ and $Y_p(x)$ satisfy the same system of first-order equations. Hence, by conditions (1.17) and (1.25),

$$\left.\frac{\partial^p R_2(x,\lambda)}{\partial\lambda^p}\right|_{\lambda=\lambda_j} = Y_p(x), \qquad p = 0,1,\dots,m. \tag{1.26}$$

Relations (1.20) and (1.26) show that the system of eigenvectors and generalized eigenvectors of the pencil $L(\lambda)$ is orthogonal to $C(x) = A^*(x) - B^*(x)$.

Let us note that the operator iSK is selfadjoint and

$$\sum_j |\mu_j|^p < \infty, \qquad p > 1, \tag{1.27}$$

where μ_j are eigenvalues of iSK. The operator SK has the property that the equality $SKY = 0$ holds only for $Y = 0$. Further, by (1.22) and (1.23), the operator T is compact. Therefore, the pencil $L(\lambda)$ satisfies all the conditions of the well-known theorem of Keldysh on the completeness of the system of eigenvectors and generalized eigenvectors. Now the assertion of the theorem is seen to be a consequence of the orthogonality of this system to $C(x)$. □

8.1.2. Let us apply Theorem 1.1 to canonical systems.

Theorem 1.2. *Suppose that there are given two canonical systems*

$$\frac{dW_k(x,z)}{dx} = izJ\mathcal{H}_k(x)W_k(x,z), \qquad 0 \le x \le l < \infty, \tag{1.28}$$

where

$$\mathcal{H}_k(x) = \mathcal{H}_k^*(x), \qquad W_k(0,z) = I_{2m}, \qquad k = 1,2. \tag{1.29}$$

Suppose further that the following equalities hold:

$$\mathcal{H}_1(0) = \mathcal{H}_2(0), \qquad W_1(l,z) = W_2(l,z), \tag{1.30}$$

$$J\mathcal{H}_k(x) = V_k(x)DV_k^{-1}(x). \tag{1.31}$$

Here the matrix D is constant and is defined by relation (1.3) *and $V_k(x)$ are continuous invertible matrices such that*

$$\|V_k'(x)\| \in L^2(0,l), \qquad k = 1,2.$$

Then

$$\mathcal{H}_1(x) = \mathcal{H}_2(x). \tag{1.32}$$

Proof. We shall consider systems (1.28) on the segment $[0, 2l]$, setting

$$\mathcal{H}_k(x+l) = \mathcal{H}_k(l-x), \qquad 0 \le x \le l. \tag{1.33}$$

Then

$$\mathcal{W}_k(2l, z) = \widetilde{\mathcal{W}}_k(l, z)\mathcal{W}_k(l, z), \qquad k = 1, 2, \tag{1.34}$$

where

$$\widetilde{\mathcal{W}}_k(l, z) = \int_l^{\curvearrowleft 2l} \exp[izJ\mathcal{H}_k(x)\,dx]. \tag{1.35}$$

Performing the change of variable $x = l + t$ we obtain

$$\widetilde{\mathcal{W}}_k(l, z) = \int_0^{\curvearrowleft l} \exp[izJ\mathcal{H}_k(l-t)\,dt] = \int_0^{\curvearrowleft l} \exp[izJ\mathcal{H}_k(x)\,dx], \tag{1.36}$$

which in conjunction with the equality

$$[J\mathcal{H}_k(t)]^* = J[J\mathcal{H}_k(t)]J$$

yields the relation

$$\widetilde{\mathcal{W}}_k(l, z) = J\mathcal{W}_k^*(l, -\bar{z})J. \tag{1.37}$$

Next, by (1.34) and (1.37),

$$\mathcal{W}_k(2l, z) = J\mathcal{W}_k^*(l, -\bar{z})J\mathcal{W}_k(l, z), \qquad k = 1, 2.$$

In view of assumption (1.30), this means that

$$\mathcal{W}_1(2l, z) = \mathcal{W}_2(2l, z). \tag{1.38}$$

Now from (1.31) and (1.33) we obtain

$$J\mathcal{H}_k(x) = \widetilde{V}_k(x)D\widetilde{V}_k^{-1}(x), \tag{1.39}$$

where

$$\widetilde{V}_k(x) = V_k(x) \cdot U_k(x), \qquad V_k(x+l) = V_k(l-x). \tag{1.40}$$

Here $U_k(x)$ are block-diagonal matrix-valued functions. Choose $U_k(x)$ so that the matrices

$$A(x) = -\widetilde{V}_1^{-1}(x)\frac{d\widetilde{V}_1}{dx}, \qquad B(x) = -\widetilde{V}_2^{-1}(x)\frac{d\widetilde{V}_2}{dx} \tag{1.41}$$

satisfy relations (1.5). By (1.30), the matrices $U_k(0)$ can be chosen so that

$$\widetilde{V}_1(0) = \widetilde{V}_2(0). \tag{1.42}$$

Now introduce the matrix-valued functions

$$W_k(x,z) = \widetilde{V}_k^{-1}(x)\mathcal{W}_k(x,z)\widetilde{V}_k(0). \tag{1.43}$$

It follows from (1.28) that

$$\frac{dW_1}{dx} = [izD + A(x)]W_1, \qquad \frac{dW_2}{dx} = [izD + B(x)]W_2. \tag{1.44}$$

By equalities (1.38), (1.42), and (1.43), we have

$$W_2(2l,z)W_1^{-1}(2l,z) = \widetilde{V}_2^{-1}(2l)\widetilde{V}_1(2l). \tag{1.45}$$

Also, by (1.40) and (1.42), the matrix $\widetilde{V}_2^{-1}(2l)\widetilde{V}_1(2l)$ is block-diagonal. Applying Theorem 1.1, we conclude that

$$A(x) = B(x). \tag{1.46}$$

Hence, by (1.41) and (1.42),

$$\widetilde{V}_1(x) = \widetilde{V}_2(x), \tag{1.47}$$

which in conjunction with (1.39) yields equality (1.32). This completes the proof of the theorem. □

8.2. Spectral data and uniqueness theorems

8.2.1. Let $\beta(x)$ be a matrix of the form

$$\beta(x) = [\beta_1(x)\ \ \beta_2(x)], \tag{2.1}$$

where $\beta_1(x)$ and $\beta_2(x)$ are $m \times m$ matrices satisfying the following conditions:

$$\beta_2(x)\beta_1^*(x) + \beta_1(x)\beta_2^*(x) = I_m, \tag{2.2}$$

$$\|\beta_k'(x)\| \leq C, \qquad 0 \leq x \leq l, \qquad k = 1, 2. \tag{2.3}$$

Further, set

$$\mathcal{H}(x) = \beta^*(x)\beta(x), \tag{2.4}$$

$$j = \begin{bmatrix} I_m & 0 \\ 0 & -I_m \end{bmatrix}. \tag{2.5}$$

Lemma 2.1. *Conditions* (2.1)–(2.3) *are equivalent to the representation*

$$J\mathcal{H}(x) = V(x)\begin{bmatrix} I_m & 0 \\ 0 & 0 \end{bmatrix} V^{-1}(x), \qquad 0 \le x \le l. \tag{2.6}$$

where $V(x)$ has a bounded derivative on the interval $[0, l]$ and satisfies the relation

$$V(x)jV^*(x) = J. \tag{2.7}$$

Proof. Let $h \neq 0$ be arbitrary $m \times 1$ column vector. Then, by condition (2.2),

$$h^*[\beta_2(x)\beta_1^*(x) + \beta_1(x)\beta_2^*(x)]h = \|h\|^2 \neq 0,$$

whence $h^*\beta_1(x) \neq 0$, i.e., $\det \beta_1(x) \neq 0$.

This allows us to express $V(x)$ and $V^{-1}(x)$ explicitly in terms of $\beta_1(x)$ and $\beta_2(x)$:

$$V(x) = \begin{bmatrix} \beta_2^*(x) & -\beta_1^{-1}(x)\beta_2(x)\beta_1^*(x) \\ \beta_1^*(x) & \beta_1^*(x) \end{bmatrix}, \tag{2.8}$$

and

$$V^{-1}(x) = \begin{bmatrix} \beta_1(x) & \beta_2(x) \\ -\beta_1(x) & \beta_1(x)\beta_2^*(x)\beta_1^{*-1}(x) \end{bmatrix}. \tag{2.9}$$

The validity of relations (2.6) and (2.7) can be verified by a straightforward calculation. The lemma is proved. □

Next, let us introduce an operator K on $L_m^2(0, l)$ as follows:

$$Kf = i\beta(x)\int_0^x \beta^*(t)f(t)\,dt, \qquad 0 \le x \le l. \tag{2.10}$$

Conditions (2.1)–(2.3) imply that K is linearly similar to the operator (see [81])

$$Af = i\int_0^x f(t)\,dt, \qquad f \in L_m^2(0, l), \tag{2.11}$$

That is to say, there exists a bounded operator $\mathcal{L}_1$ with bounded inverse such that

$$K = \mathcal{L}_1^{-1} A \mathcal{L}_1.$$

Consider the operator

$$\mathcal{L} = (I + R)\mathcal{L}_1, \tag{2.12}$$

where R acts according to the rule

$$Rf = \int_0^x r(x-t)f(t)\,dt, \qquad f \in L_m^2(0, l). \tag{2.13}$$

Since $AR = RA$, it follows that

$$K = \mathcal{L}^{-1} A \mathcal{L}. \tag{2.14}$$

We can choose the matrix-valued function $r(x)$ in such a manner that

$$\mathcal{L}\beta_2(x) = I_m. \tag{2.15}$$

This means that $\mathcal{L}$ maps the columns of the matrix $\beta_2(x)$ into the corresponding columns of I_m.

Then the identity

$$(K - K^*)f = i\beta(x)J \int_0^l \beta^*(t)f(t)\,dt \tag{2.16}$$

and relations (2.1), (2.14), and (2.15) yield the operator identity

$$(AS - SA^*)f = i \int_0^l [M(x) + M^*(t)]f(t)\,dt, \tag{2.17}$$

where

$$S = \mathcal{L}\mathcal{L}^* \qquad M(x) = \mathcal{L}\beta_1(x). \tag{2.18}$$

Relation (2.17) can be recast as the S-node

$$AS - SA^* = i(\Phi_1\Phi_2^* + \Phi_2\Phi_1^*), \tag{2.19}$$

where

$$\Phi_1 g = M(x)g, \qquad \Phi_2 g = g. \tag{2.20}$$

Here g are the constant $m \times 1$ vectors.

Let us prove the following result.

Theorem 2.1. *Assume that conditions* (2.1)–(2.4) *are satisfied. Then the canonical system*

$$\frac{d\mathcal{W}(x,z)}{dx} = izJ\mathcal{H}(x)\mathcal{W}(x,z) \tag{2.21}$$

is uniquely determined by its spectral data $\tau(u)$ *and* α.

Proof. The operator $\mathcal{L}$ has the form [81]

$$\mathcal{L}f = f(x) + \int_0^x \mathcal{L}(x,t)f(t)\,dt, \qquad f \in L_m^2(0,l). \tag{2.22}$$

By formula (2.7)of Chapter 4, the canonical system

$$\frac{d\mathcal{W}(x,z)}{dx} = izJ\widetilde{\mathcal{H}}(x)\mathcal{W}(x,z) \tag{2.23}$$

is associated with the operator identity (2.18). In (2.23) $\widetilde{\mathcal{H}}(\zeta)$ is defined by the formula

$$\widetilde{\mathcal{H}}(\zeta) = \frac{d}{d\zeta}\left(\Pi^* S_\zeta^{-1} P_\zeta \Pi\right), \qquad \Pi = [\Phi_1 \ \ \Phi_2], \tag{2.24}$$

where

$$P_\zeta f = \begin{cases} f, & \text{for } 0 \le x \le \zeta \\ 0 & \text{for } \zeta < x. \end{cases} \tag{2.25}$$

From (2.22), (2.24), and (2.25) it follows that

$$\widetilde{H}(x) = \left\{(\mathcal{L}^{-1}\Phi_j)^* \cdot (\mathcal{L}^{-1}\Phi_i)\right\}_{i,j=1}^2 . \tag{2.26}$$

Finally, by (2.14), (2.17) and (2.18),

$$\beta_1(x) = \mathcal{L}^{-1}\Phi_1, \qquad \beta_2(x) = \mathcal{L}^{-1}\Phi_2,$$

whence $\mathcal{H}(x) = \widetilde{\mathcal{H}}(x)$. This completes the proof of the theorem. □

8.2.2. Assume again that the matrix $\beta(x)$ is of the form (2.1), and let the relations

$$\beta(x)J\beta^*(x) = 0, \qquad i\beta(x)J[\beta'(x)]^* = I_m \tag{2.27}$$

be satisfied. In addition, assume that there exists a constant $C > 0$ such that

$$\|\beta_k''(x)\| \le C, \qquad 0 \le x \le l, \qquad k = 1, 2. \tag{2.28}$$

Lemma 2.2. *Let $\mathcal{H}(x)$ be defined by (2.4). Conditions (2.27) and (2.28) are equivalent to the representation*

$$J\mathcal{H}(x) = V(x)\begin{bmatrix} 0 & 0 \\ I_m & 0 \end{bmatrix} V^{-1}(x), \tag{2.29}$$

where the matrix $V(x)$ and its blocks V_{ij} satisfy the relations

$$V(x)JV^*(x) = J, \tag{2.30}$$

$$i[V_{12}^*(x)V_{22}'(x) + V_{22}^*(x)V_{12}'(x)] = I_m, \tag{2.31}$$

and the blocks $V_{12}(x)$, $V_{22}(x)$ have bounded second derivatives.

Proof. It can be readily verified that the matrix

$$V(x) = \begin{bmatrix} i\beta_2^{*\prime}(x) & \beta_2^*(x) \\ i\beta_1^{\prime *}(x) & \beta_1^*(x) \end{bmatrix} \cdot \begin{bmatrix} I_m & 0 \\ -\frac{1}{2}\gamma(x) & I_m \end{bmatrix}, \tag{2.32}$$

where

$$\gamma(x) = \beta_1{}'(x)\beta_2^{*\prime}(x) + \beta_2{}'(x)\beta_1^{*\prime}(x),$$

satisfies conditions (2.29)–(2.31). □

8.2.3. Consider again the operator K defined by (2.10). Conditions (2.27) and (2.28) imply that K is now similar to the operator A acting as

$$Af = \int_0^x (t-x)f(t)\,dt, \qquad 0 \le x \le l, \tag{2.33}$$

i.e., $K = \mathcal{L}_1^{-1} A \mathcal{L}_1$ (see [81]). Assume also that

$$\det \beta_2(0) \neq 0. \tag{2.34}$$

Let us introduce the operator

$$\mathcal{L} = [\beta_2(0) + R]\mathcal{L}_1, \tag{2.35}$$

where R is of the form (2.13). We can choose $r(x)$ so that (2.15) is satisfied. Then, by (2.14) and (2.16), relations (2.17) and (2.18) hold. The following theorem can be established in much the same manner as Theorem 2.1.

Theorem 2.2. *Assume that conditions* (2.27), (2.28) *and* (2.34) *are satisfied. Then the canonical system* (2.4), (2.21) *is uniquely determined by its spectral data* $\tau(u)$ *and* α.

Chapter 9
Weyl Discs and Points

In this chapter we consider the canonical system

$$\frac{d\mathcal{W}}{dx} = izJ\mathcal{H}(x)\mathcal{W}(x,z), \qquad \mathcal{H}(x) \geq 0, \qquad 0 \leq x < \infty, \tag{0.1}$$

together with the Hilbert space $L^2(\mathcal{H})$ with the inner product

$$(Y_1, Y_2) = \int_0^\infty Y_2^*(x)\mathcal{H}(x)Y_1(x)\,dx, \tag{0.2}$$

where $Y_1(x)$ and $Y_2(x)$ are $2m \times 1$ column vectors. In Section 1 we investigate the set of the matrix-valued functions $v(l,z)$ of the form

$$v(l,z) = i[a(l,z)R(z) + b(l,z)Q(z)] \cdot [c(l,z)R(z) + d(l,z)Q(z)]^{-1}, \tag{0.3}$$

where the matrix of coefficients is given by

$$\mathfrak{A}(l,z) = \mathcal{W}^*(l,\bar{z}) = \begin{bmatrix} a(l,z) & b(l,z) \\ c(l,z) & d(l,z) \end{bmatrix} \tag{0.4}$$

and the pair $R(z)$, $Q(z)$ is non-singular and has the J-property.

The matrices $v(l,z)$ are analogues of the classical Weyl-Titchmarsh matrices. As in the classical case, they form a system of nested *Weyl discs.* In Section 2 we derive conditions under which the limit Weyl disc ($l \to \infty$) reduces to a point, the *Weyl limit point*; in this case we say that the given canonical system is in the *Weyl limit-point case.* There we prove theorems that generalize the well-known result of B. M. Levitan [57], obtained for Dirac-type systems. Let us note that formulas (0.3) and (0.4) were introduced in Chapter 4 in connection with the spectral characteristics of the system (0.1). In this chapter they appear in connection with the problem of the number of linearly independent solutions of system (0.1) that belong to the space $L^2(\mathcal{H})$.

In Sections 3 and 4 we consider the canonical system (0.1) on the line $-\infty < x < \infty$ and investigate the corresponding Weyl-Titchmarsh matrix-valued function $v(z)$.

9.1. Basic notions

9.1.1. Let us denote by $\mathcal{K}_l$ the set of the matrix-valued functions described by formulas (0.3), (0.4). Since $R(z)$, $Q(z)$ have the J-property, the following assertion holds true.

Proposition 1.1. *Let $v(z)$ be an analytic matrix-valued function in the upper half-plane with the property that $i^{-1}[v(z) - v^*(z)] \geq 0$ for $\operatorname{Im} z > 0$. Then $v(z)$ belongs to $\mathcal{K}_l$ if and only if the following inequality holds:*

$$[iv^*(z) \;\; I_m]\mathfrak{A}^{*-1}(l,z)J\mathfrak{A}^{-1}(l,z)\begin{bmatrix} -iv^*(z) \\ I_m \end{bmatrix} \geq 0. \tag{1.1}$$

Proposition 1.2. *The J-form*

$$J - \mathfrak{A}^{-1*}(l,z)J\mathfrak{A}^{-1}(l,z),$$

of the matrix $\mathfrak{A}^{-1}(l,z)$ increases when the parameter l increases and $\operatorname{Im} z > 0$.

Proof. Using the system (0.1), we can write

$$\mathfrak{A}^{-1}(l,z) = \overset{\curvearrowleft}{\int_0^l} \exp[iz\mathcal{H}(x)J\,dx]. \tag{1.2}$$

Consequently,

$$\mathfrak{A}^{-1}(l_2,z) = V(z)\mathfrak{A}^{-1}(l_1,z), \qquad l_2 > l_1, \tag{1.3}$$

where

$$V(z) = \overset{\curvearrowleft}{\int_{l_1}^{l_2}} \exp[iz\mathcal{H}(x)J\,dx]. \tag{1.4}$$

It follows from relations (1.2)–(1.4) that $\mathfrak{A}^{-1}(l_1,z)$ and $V(z)$ are J-contractive matrices. Hence, taking the difference of the J-forms for $\mathfrak{A}^{-1}(l_2,z)$ and $\mathfrak{A}^{-1}(l_1,z)$, we have

$$\begin{aligned}&[J - \mathfrak{A}^{*-1}(l_2,z)J\mathfrak{A}^{-1}(l_2,z)] - [J - \mathfrak{A}^{-1*}(l_1,z)J\mathfrak{A}^{-1}(l_1,z)]\\ &=\mathfrak{A}^{-1*}(l_1,z)[J - V^*(z)JV(z)]\mathfrak{A}^{-1}(l_1,z) \geq 0\end{aligned} \tag{1.5}$$

The proposition is proved. □

Proposition 1.2 shows that the *Weyl matrices* (see [65])

$$M(l,z) = \mathfrak{A}^{*-1}(l,z)J\mathfrak{A}^{-1}(l,z) = \begin{bmatrix} -r(l,z) & q(l,z) \\ q^*(l,z) & -t(l,z) \end{bmatrix}$$

are monotonically decreasing in l. Moreover, since $M(l,z) \leq J$, one has that $r(l,z) \geq 0$. We will assume that actually the strict inequality holds:

$$r(l,z) > 0 \qquad l > l_0. \tag{1.6}$$

It follows from (1.1) and (1.5) that

$$v^*(z)r(l,z)v(z) + iq^*(l,z)v(z) - iv^*(z)q(l,z) \leq -t(l,z). \tag{1.7}$$

Taking into account (1.6), we can rewrite the last inequality in the form

$$[-iv^*(z)r^{1/2}(l,z)+q^*(l,z)r^{-1/2}(l,z)][r^{1/2}(l,z)iv(z)+r^{-1/2}(l,z)q(l,z)] \le q^*(l,z)r^{-1}(l,z)q(l,z)-t(l,z). \tag{1.8}$$

This shows that

$$\rho_1(l,z)=q^*(l,z)r^{-1}(l,z)q(l,z)-t(l,z)\ge 0. \tag{1.9}$$

Hence, if we introduce T via the relation

$$r^{1/2}(l,z)iv(z)+r^{-1/2}(l,z)q(l,z)=T\rho_1^{1/2}(l,z), \tag{1.10}$$

then (1.8) becomes $T^*T\le I$. From (1.10) we obtain the representation

$$v(z)=\Gamma(l,z)+\rho_2^{1/2}(l,z)T\rho_1^{1/2}(l,z), \tag{1.11}$$

where

$$\Gamma(l,z)=ir^{-1}(l,z)q(l,z), \qquad T^*T\le I, \tag{1.12}$$

and

$$\rho_2(l,z)=r^{-1}(l,z). \tag{1.13}$$

The set of matrices admitting representation (1.11)–(1.13) is called a *Weyl matrix disc*. Thus, inequality (1.1) has the following geometric interpretation [63].

Theorem 1.1. *The matrix-valued functions $v(z)$ satisfying* (1.1) *for a fixed z form a matrix disc $\mathcal{K}_l$ with the centre*

$$\Gamma(l,z)=ir^{-1}(l,z)q(l,z), \tag{1.14}$$

the right radius

$$\rho_1(l,z)=q^*(l,z)r^{-1}(l,z)q(l,z)-t(l,z), \tag{1.15}$$

and the left radius

$$\rho_2(l,z)=r^{-1}(l,z). \tag{1.16}$$

9.1.2. Let us examine the properties of the Weyl discs $\mathcal{K}_l$. Since $M(l,z)$ is monotonically decreasing (Proposition 1.2),

$$[-iv^*(z)\ \ I_m]M(l_1,z)\begin{bmatrix} iv(z)\\ I_m\end{bmatrix} \ge [-iv^*(z)\ \ I_m]M(l_2,z)\begin{bmatrix} iv(z)\\ I_m\end{bmatrix}, \qquad l_2>l_1.$$

Consequently, $\mathcal{K}_{l_2}\subset\mathcal{K}_{l_1}$ for $l_2>l_1$. The following properties are evident [65].

I. The left and the right radii of the Weyl discs are monotonically decreasing functions of l and therefore when $l\to\infty$ they have limits

$$\rho_1(z)=\lim_{l\to\infty}\rho_1(l,z), \qquad \rho_2(z)=\lim_{l\to\infty}\rho_2(l,z). \tag{1.17}$$

II. The family of centres $\Gamma(l,z)$ is bounded.

By a theorem of S. A. Orlov [63], the following assertion holds true.

III. The ranks of the radii $\rho_1(z)$ are $\rho_2(z)$ do not depend on the choice of z ($\operatorname{Im} z > 0$).

If the family of centres $\Gamma(l,z)$ has a limit

$$\Gamma(z) = \lim_{l\to\infty} \Gamma(l,z), \tag{1.18}$$

then the set formed by the matrices

$$v(z) = \Gamma(z) + \rho_2^{1/2}(z)T\rho_1^{1/2}(z), \qquad T^*T \le I, \tag{1.19}$$

is called the *Weyl limit disc*, denoted here $\mathcal{K}$. If at least one of the limit radii $\rho_1(z)$ or $\rho_2(z)$ is zero, the Weyl limit disc reduces to a point, called the *Weyl limit point.* In this case we say that the canonical system (0.1) is in the *Weyl limit-point case.*

Remark 1.1. Suppose that

$$\int_0^l \mathcal{H}(x)\,dx > 0\,, \tag{1.20}$$

i.e., system (0.1) is of positive type. Then condition (1.6) satisfied.

9.1.3. Consider the pair

$$\begin{bmatrix} u_1(z) \\ u_2(z) \end{bmatrix} = \mathfrak{A}(l,z) \begin{bmatrix} R(z) \\ Q(z) \end{bmatrix}. \tag{1.21}$$

Proposition 1.4. *If $R(z)$, $Q(z)$ is a non-singular pair then*

$$\det u_2(z) \not\equiv 0, \qquad \operatorname{Im} z > 0.$$

Proof. By definition (1.21),

$$\begin{aligned}&([R^*(z)Q(z) + Q^*(z)R(z)]e,\, e) = \\ &= ([u_1^*(z)\ u_2^*(z)])\mathfrak{A}^{*-1}(l,z)J\mathfrak{A}^{-1}(l,z)\begin{bmatrix} u_1(z) \\ u_2(z) \end{bmatrix} e,\, e).\end{aligned} \tag{1.22}$$

Suppose that there is a vector $e \ne 0$ such that

$$u_2(z)e = 0. \tag{1.23}$$

Then, by (1.22),

$$([R^*(z)Q(z) + Q^*(z)R(z)]e,\, e) = (u_1^*(z)r(l,z)u_1(z)e, e)\,,$$

which in conjunction with (1.6) implies that

$$u_1(z)e = u_2(z)e = 0.$$

This contradicts the nonsingularity of the pair $R(z)$, $Q(z)$, and completes the proof of the proposition. □

Proposition 1.5. *Let $v(z)$ belong to the Weyl limit disc of the system* (0.1). *Then the following inequality holds*

$$\int_0^\infty [I_m \;\; iv^*(z)]\mathcal{W}^*(x,z)\mathcal{H}(x)\mathcal{W}(x,z)\begin{bmatrix} I_m \\ -iv(z)\end{bmatrix} dx < \infty. \tag{1.24}$$

Proof. Let the pairs $R(l,z)$, $Q(l,z)$ be such that $v(z) = \lim v(l,z)$ when $l \to \infty$. From (1.21) it follows that the columns of the matrices

$$\begin{bmatrix} -iv(l,z) \\ I_m \end{bmatrix}, \qquad \mathcal{W}^*(l,\bar{z})\begin{bmatrix} R \\ Q \end{bmatrix}$$

span the same linear subspace. Hence, thanks to the equality

$$\mathcal{W}^*(l,\bar{z})J\mathcal{W}(l,z) = J, \tag{1.25}$$

the columns of the matrices $J\mathcal{W}(l,z)\operatorname{col}[I_m, -iv(l,z)]$ and $\operatorname{col}[R,Q]$ span the same subspace. By the J-property of the pair R, Q, this subspace is J-nonnegative. Consequently,

$$[I_m \;\; iv^*(l,z)]\mathcal{W}^*(l,z)J\mathcal{W}(l,z)\begin{bmatrix} I_m \\ -iv(l,z)\end{bmatrix} \geq 0 \tag{1.26}$$

whenever $\operatorname{Im} z > 0$. Using the relation

$$\int_0^l \mathcal{W}^*(x,z)\mathcal{H}(x)\mathcal{W}(x,z)\,dx = \frac{i}{z-\bar{z}}(J - W^*(l,z)JW(l,z)) \tag{1.27}$$

and (1.26), we deduce the inequality

$$\int_0^l [I_m \;\; iv^*(l,z)]\mathcal{W}^*(x,z)\mathcal{H}(x)\mathcal{W}(x,z)\begin{bmatrix} I_m \\ -iv(l,z)\end{bmatrix} dx \leq \frac{v(l,z)-v^*(l,z)}{z-\bar{z}}. \tag{1.28}$$

Letting here $l \to \infty$, we get

$$\int_0^l [I_m \;\; iv^*(z)]\mathcal{W}^*(x,z)\mathcal{H}(x)\mathcal{W}(x,z)\begin{bmatrix} I_m \\ -iv(z)\end{bmatrix} dx \leq \frac{v(z)-v^*(z)}{z-\bar{z}}. \tag{1.29}$$

The proposition is proved. □

Remark 1.1. The classical Weyl-Titchmarsh functions in the theory of second-order differential equations allow one construct solutions that belong to the space $L^2(0,\infty)$. It follows from (1.27) that

$$Y(x,z) = \mathcal{W}(x,z)\begin{bmatrix} I_m \\ -iv(z)\end{bmatrix}, \qquad 0 \leq x \leq \infty, \tag{1.30}$$

is a solution of the system (0.1) belonging to $L^2(\mathcal{H})$, i.e., $v(z)$ is indeed an analogue of the classical Weyl-Titchmarsh function.

9.2. Symmetric operators and deficiency indices

9.2.1. Let A be a linear operator with dense domain $\mathfrak{D}(A)$ in $\mathfrak{H}$. The operator A is said to be *symmetric*

$$(Ax, y) = (x, Ay) \tag{2.1}$$

for all $x, y \in \mathfrak{D}(A)$.

Let us list the results of the theory of symmetric operators that are needed to continue our investigation.

Symmetric operators have the property that

$$A \subset A^*. \tag{2.2}$$

Denote by $\mathfrak{L}_\lambda$ the linear manifold

$$\{y : y = (A - \lambda I)x, \ x \in \mathfrak{D}(A)\}, \tag{2.3}$$

and by $\mathfrak{N}_\lambda$ the subspace orthogonal to $\mathfrak{L}_\lambda$.

The numbers m_+, m_-, where

$$m_+ = \dim \mathfrak{N}_\lambda \quad (\operatorname{Im}\lambda > 0), \qquad m_- = \dim \mathfrak{N}_\lambda \quad (\operatorname{Im}\lambda < 0) \tag{2.4}$$

are called the *deficiency indices* of the symmetric operator A.

Here we used the fact that, in each of the half-planes ($\operatorname{Im}\lambda > 0$) and ($\operatorname{Im}\lambda < 0$), $\dim \mathfrak{N}_l$ does not depend on λ [63].

Moreover, the following *von Neumann formula* holds true:

$$\mathfrak{D}(A^*) = \mathfrak{D}(A) \oplus \mathfrak{N}_{\bar{\lambda}} \oplus \mathfrak{N}_\lambda. \tag{2.5}$$

9.2.2. In the space $L^2(\mathcal{H})$, equipped with the inner product

$$(Y_1, Y_2) = \int_0^\infty Y_2^*(x)\mathcal{H}(x)Y_1(x)\,dx, \tag{2.6}$$

consider the operator

$$L_0 Y = -i\mathcal{H}^{-1}(x)J\frac{dY}{dx}, \qquad Y(0) = 0, \tag{2.7}$$

where $Y(x)$ are compactly supported absolutely continuous vector-valued functions.

We will assume that the matrix-valued function $H(x)$ is locally integrable and

$$\mathcal{H}(x) > 0, \qquad 0 \le x < \infty. \tag{2.8}$$

Then the operator L_0 is symmetric.

The following assertion is a direct consequence of definition (2.4).

Proposition 2.1. *The number of the linearly independent solutions of the system*

$$\frac{dY}{dx} = i\bar{l}J\mathcal{H}(x)Y, \qquad 0 \le x < \infty \tag{2.9}$$

that belong to $L^2(\mathcal{H})$ *coincides with* m_+ *when* $\operatorname{Im}\lambda > 0$, *and with* m_- *when* $\operatorname{Im}\lambda < 0$.

The exact values of the deficiency indices of the operator L_0 can be calculated if we make additional assumptions.

Theorem 2.1. *Let the matrix-valued function* $\mathcal{H}(x)$ *be continuous and let the inequality*

$$\mathcal{H}(x) \ge \delta I_{2m}, \qquad 0 \le x < \infty \tag{2.10}$$

hold for some constant $\delta > 0$. *Then*

$$m_+ = m_- = m. \tag{2.11}$$

Proof. Write $Y(x)$ in the block form

$$Y(x) = \begin{bmatrix} y_1(x) \\ y_2(x) \end{bmatrix},$$

where $y_k(x)$ $(k = 1, 2)$ are $m \times 1$ column vectors. Consider the extension $\widetilde{L}$ of the operator L_0 defined by

$$\widetilde{L}_0 Y = -i\mathcal{H}^{-1}(x)J\frac{dY}{dx}, \qquad y_1(0) = 0. \tag{2.12}$$

If $Y(x), Z(x) \in \mathfrak{D}(L_0^*)$, then

$$y_1(0) = z_1(0) = 0, \tag{2.13}$$

and

$$Y(x), Z(x) \in L^2(\mathcal{H}). \tag{2.14}$$

It follows from (2.14) that there exists a sequence $a_n \to \infty$ such that

$$Z^*(a_n)JY(a_n) \to 0. \tag{2.15}$$

Using (2.13) and (2.15), we calculate the inner product

$$(\widetilde{L}_0^* Y, Z) = \lim_{n\to\infty} \int_0^{a_n} Z^*(x)\left(-iJ\frac{dY}{dx}\right) dx = \int_0^{\infty} \frac{dZ^*}{dx}(iJY)\,dx = (Y, \widetilde{L}_0^* Z).$$

This means that the operator $\widetilde{L}_0$ is selfadjoint, and consequently its deficiency indices $\widetilde{m}_+$ and $\widetilde{m}_-$ are both equal to zero. This in turn implies that the deficiency indices of the operator L_0 satisfy relations (2.11). The theorem is proved. □

In a number of cases the requirement (2.10) proves to be too strong. Repeating the arguments of Theorem 2.1, one can establish the following assertion.

Theorem 2.2. *Let the matrix $\mathcal{H}(x)$ have the form*

$$\mathcal{H}(x) = T^*(x)\mathcal{H}_1(x)T(x) \tag{2.16}$$

and the following conditions be satisfied:

1) *The matrices $T(x)$ and $\mathcal{H}_1(x)$ are continuous.*
2) *The inequality*
$$\mathcal{H}_1(x) \geq \delta I_{2m}, \qquad 0 \leq x < \infty \tag{2.17}$$
holds for some $\delta > 0$.
3) *There exists a continuous matrix-valued function $Q(x)$, bounded on the half-line $0 \leq x < \infty$, such that*
$$T^*(x)Q(x)T(x) = J. \tag{2.18}$$

Then the deficiency indices of the operator L_0 are given by relations (2.11).

Proof. Hypotheses (2.16) and (2.17) guarantee that

$$T(a_n)Y(a_n) \to 0, \qquad T(a_n)Z(a_n) \to 0$$

as $n \to \infty$, for arbitrary $Y(x), Z(x) \in L^2(\mathcal{H})$. Now hypothesis 3) implies that

$$Z^*(a_n)T^*(a_n)Q(a_n)T(a_n)Y(a_n) = Z^*(a_n)JY(a_n) \to 0.$$

It remains to argue as in Theorem 2.1. □

Remark 2.1. Suppose that, in addition to the conditions of Theorem 2.2, the following requirements are fulfilled:

$$\mathcal{H}_1(x) = I_{2m}, \qquad Q(x) = J,$$

and the matrix-valued function $T(x)$ is continuously differentiable. Then Theorem 2.2 coincides with a celebrated theorem of B. M. Levitan [57]. In fact, our proof is similar to Levitan's proof.

From relations (1.22) and (2.11) we deduce the following assertion.

Corollary 2.1. *Let the conditions of either Theorem 2.1 or Theorem 2.2 be satisfied. Then the corresponding canonical system (0.1) on the half-line is in the Weyl limit-point case (i.e., the Weyl limit disc reduces to a point).*

9.3. Weyl-Titchmarsh matrix functions on the line

9.3.1. Let $B(x)$ be a continuous monotonically increasing $m \times m$ matrix-valued function. Consider the system

$$\mathcal{W}(x,z) = I_{2m} + izJ \int_0^x [dB(t)]\mathcal{W}(t,z), \qquad -\infty < x < \infty. \tag{3.1}$$

Let us replace this system by the doubled system on the half-line:

$$\mathcal{W}_0(x,z) = I_{4m} + izJ_0 \int_0^x [dB_0(t)]\mathcal{W}_0(t,z), \qquad 0 \le x < \infty. \tag{3.2}$$

Here the matrices J_0, $B_0(x)$, and $\mathcal{W}_0(x,z)$ are defined by formulas

$$\mathcal{W}_+(x,z) = \mathcal{W}(x,z), \quad \mathcal{W}_-(x,z) = \mathcal{W}(-x,z), \quad x \ge 0, \tag{3.3}$$

$$B_+(x) = B(x), \quad B_-(x) = -B(-x), \quad x \ge 0, \tag{3.4}$$

$$\mathcal{W}_0(x,z) = T^{-1} \begin{bmatrix} \mathcal{W}_+(x,z) & 0 \\ 0 & \mathcal{W}_-(x,z) \end{bmatrix} T, \tag{3.5}$$

$$B_0(x) = T^{-1} \begin{bmatrix} B_+(x) & 0 \\ 0 & B_-(x) \end{bmatrix} T, \tag{3.6}$$

$$T = \frac{1}{\sqrt{2}} \begin{bmatrix} J & I_{2m} \\ -J & I_{2m} \end{bmatrix}, \qquad J_0 = \begin{bmatrix} 0 & I_{2m} \\ I_{2m} & 0 \end{bmatrix}. \tag{3.7}$$

By Theorem 1.1 of Chapter 5, the Weyl-Titchmarsh matrix functions of the system (3.1) coincide with the Weyl-Titchmarsh matrix functions of the system (3.2). Let us introduce the systems

$$\mathcal{W}_\pm(x,z) = I_{2m} \pm izJ \int_0^x [dB_\pm(t)]\mathcal{W}_\pm(t,z), \qquad 0 \le x < \infty. \tag{3.8}$$

The Weyl-Titchmarsh matrix functions $v(z)$ and $v_\pm(z)$ of systems (3.2) and (3.8) are defined by the inequalities

$$\int_0^\infty [I_{2m} \;\; iv^*(z)]\mathcal{W}_0^*(x,z)[dB_0(x)]\mathcal{W}_0(x,z) \begin{bmatrix} I_{2m} \\ -iv(z) \end{bmatrix} < \infty, \quad \text{Im } z > 0, \tag{3.9}$$

and

$$\int_0^\infty [I_m \;\; \pm iv_\pm^*(z)]\mathcal{W}_\pm^*(x,z)[dB_\pm(x)]\mathcal{W}_\pm(x,z) \begin{bmatrix} I_m \\ \mp iv_\pm(z) \end{bmatrix} < \infty, \quad \text{Im } z > 0, \tag{3.10}$$

respectively. Here we used the fact that Proposition 1.5 holds true for system (3.2) as well as for system (3.8). By (3.5) and (3.7), inequality (3.9) is equivalent to the inequalities

$$\int_0^\infty (J \pm iv^*(z))\mathcal{W}_\pm^*(x,z)[dB_\pm(x)]\mathcal{W}_\pm(x,z)(J \mp iv(z)) < \infty. \tag{3.11}$$

Now let us write $v(z)$ in block form

$$v(z) = \{v_{ij}(z)\}_{ij=1}^2$$

where $v_{ij}(z)$ are $m \times m$ matrices.

Theorem 3.1. *Suppose that unique Weyl-Titchmarsh matrix-valued functions $v_-(z)$ and $v_+(z)$ correspond to problems (3.8) when $l = \infty$ (i.e., we are in the Weyl limit-point case). If the relations*

$$\det v_{11}(z) \not\equiv 0, \quad \det[I_m \pm iv_{12}(z)] \not\equiv 0, \quad \operatorname{Im} z > 0 \tag{3.12}$$

hold, then

$$v_+(z) = -[I_m - iv_{21}(z)]v_{11}^{-1}(z) = v_{22}(z)[I_m - iv_{12}(z)]^{-1} \tag{3.13}$$

$$v_-(z) = -[I_m + iv_{21}(z)]v_{11}^{-1}(z) = v_{22}(z)[I_m + iv_{12}(z)]^{-1}, \tag{3.14}$$

and

$$[Jv(z)]^2 = -I_{2m}. \tag{3.15}$$

Proof. Relations (3.11) can be rewritten in the equivalent form

$$\int_0^\infty [\pm iv_{11}^*(z) \;\; I_m \pm iv_{21}^*(z)]\, \mathcal{W}_\pm^*(x,z)\, [dB_\pm(x)]\, \mathcal{W}_\pm(x,z) \begin{bmatrix} \mp iv_{11}(z) \\ I_m \mp iv_{21}(z) \end{bmatrix} < \infty \tag{3.16}$$

and

$$\int_0^\infty [I_m \pm iv_{12}^*(z) \;\; \pm iv_{22}^*(z)]\, \mathcal{W}_\pm^*(x,z)\, [dB_\pm(x)]\, \mathcal{W}_\pm(x,z) \begin{bmatrix} I_m \mp iv_{12}(z) \\ \mp iv_{22}(z) \end{bmatrix} < \infty. \tag{3.17}$$

Since, by the hypotheses of the theorem, problems (3.8) are in the Weyl limit-point case, equalities (3.13) and (3.14) follow from (3.10), (3.16) and (3.17). Further, (3.13) and (3.14) imply that

$$v_{21}v_{11}^{-1} + v_{11}^{-1}v_{12} = 0, \qquad v_{22} = v_{21}v_{11}^{-1}v_{12} - v_{11}^{-1}, \tag{3.18}$$

which in turn is equivalent to equality (3.15). This completes the proof of the theorem. □

Corollary 3.1. *Let the conditions of Theorem 3.1 be satisfied. If the entries of the matrix $v(z)$ are rational functions, then so are the entries of the matrices $v_\pm(z)$.*

Let us use relations (3.13) and (3.14) to express the blocks $v_{ij}(z)$ of the matrix $v(z)$ through $v_\pm(z)$:

$$v_{11}(z) = -2[v_+(z) + v_-(z)]^{-1}, \tag{3.19}$$

$$v_{21}(z) = i[v_+(z) - v_-(z)][v_+(z) + v_-(z)]^{-1}, \tag{3.20}$$

$$v_{12}(z) = -i[v_+(z) + v_-(z)]^{-1}[v_+(z) - v_-(z)]\,, \tag{3.21}$$

and

$$v_{22}(z) = \frac{1}{2}\left\{[v_+(z) + v_-(z)] - [v_+(z) - v_-(z)][v_+(z) + v_-(z)]^{-1}[v_+(z) - v_-(z)]\right\}. \tag{3.22}$$

Corollary 3.2. *Suppose that unique Weyl-Titchmarsh functions $v_+(z)$ and $v_-(z)$ correspond to problems* (3.8) *(i.e., we are in the Weyl limit-point case). If the relations*

$$\det v_\pm(z) \not\equiv 0, \quad \det[v_+(z) + v_-(z)] \not\equiv 0, \quad \operatorname{Im} z > 0 \tag{3.23}$$

hold, then the matrix $v(z)$ whose blocks are defined by formulas (3.19)–(3.22) *is a Weyl-Titchmarsh matrix function of the system* (3.1).

9.3.2. Let us clarify the structure of the matrix-valued function

$$\tilde{v}(z) = i\left\{u_2\left[-iv(z)\right] + u_1\right\}\left\{u_1\left[-iv(z)\right] + u_2\right\}^{-1}, \tag{3.24}$$

where

$$u_1 = \begin{bmatrix} I_m & 0 \\ 0 & 0 \end{bmatrix}, \qquad u_2 = \begin{bmatrix} 0 & 0 \\ 0 & I_m \end{bmatrix}. \tag{3.25}$$

The matrix $\tilde{v}(z)$ plays an essential role in solving the inverse spectral problem on the line (see Chapter 5). Formula (3.7) can be rewritten in the form

$$\tilde{v}(z) = \begin{bmatrix} -v_{11}^{-1} & -iv_{11}^{-1}v_{12} \\ iv_{21}v_{11}^{-1} & v_{22} - v_{21}v_{11}^{-1}v_{12} \end{bmatrix}. \tag{3.26}$$

Thus, by (3.18), $\tilde{v}(z)$ has the following block structure:

$$\tilde{v}(z) = \begin{bmatrix} \tilde{v}_{11}(z) & \tilde{v}_{12}(z) \\ \tilde{v}_{12}(z) & \tilde{v}_{11}(z) \end{bmatrix}. \tag{3.27}$$

9.4. Weyl-Titchmarsh matrix function of a system with shifted argument

9.4.2. Let us consider a canonical system of the form

$$\frac{dY}{dx} = izJ\mathcal{H}(x+\sigma)Y, \qquad \sigma = \bar{\sigma}, \qquad -\infty < x < \infty, \tag{4.1}$$

where $\mathcal{H}(x)$ is a continuous matrix-valued function and

$$\mathcal{H}(x) \geq 0. \tag{4.2}$$

In this section we exhibit a connection between the Weyl-Titchmarsh function $v(\sigma, z)$ of the system (4.1) and $v(z) = v(0, z)$. The analogous problem for the Sturm-Liouville equation was already solved (see [56]). This result proved to be useful in constructing so-called *finite-band* or *finite-gap solutions.*

9.4.3. Let us introduce the following notation:

$$\mathcal{H}_\pm(x,\sigma) = \mathcal{H}(x \pm \sigma), \qquad x \geq 0. \tag{4.3}$$

The matrix-valued functions $\mathcal{W}_\pm(x,z)$ are defined to be the solutions of the problems

$$\frac{d\mathcal{W}_\pm(x,\sigma)}{dx} = \pm izJ\mathcal{H}_\pm(x,\sigma)\mathcal{W}_\pm(x,\sigma,z), \qquad \mathcal{W}_\pm(0,\sigma,z) = I_{2m}. \tag{4.4}$$

By (4.4), we can write

$$\mathcal{W}_+(x,\sigma,z) = \overset{\curvearrowleft}{\int_0^x} \exp[izJ\mathcal{H}(x+\sigma)dx]. \tag{4.5}$$

Now let us introduce the matrix-valued function

$$A(\sigma,z) = \overset{\curvearrowleft}{\int_0^\sigma} \exp[izJ\mathcal{H}(x)dx]. \tag{4.6}$$

By (4.5) and (4.6),

$$\mathcal{W}_+(x,\sigma,z) = \mathcal{W}_1(x+\sigma,z)A(\sigma,z)^{-1}, \qquad x+\sigma > 0, \tag{4.7}$$

where

$$\mathcal{W}_1(x,z) = \mathcal{W}_+(x,0,z). \tag{4.8}$$

In a similar manner one can derive the relation

$$\mathcal{W}_-(x,\sigma,z) = \mathcal{W}_2(x-\sigma,z)A(\sigma,z)^{-1}, \qquad x-\sigma > 0, \tag{4.9}$$

where

$$\mathcal{W}_2(x,z) = \mathcal{W}_-(x,0,z). \tag{4.10}$$

The Weyl-Titchmarsh functions $v(\sigma,z)$ of the system (4.1) are defined by the inequalities

$$\int_0^\infty (J \pm iv^*(\sigma,z))\mathcal{W}_\pm^*(x,\sigma,z)\mathcal{H}_\pm(x,\sigma)\mathcal{W}_\pm(x,\sigma,z)(J \mp iv(\sigma,z))\,dx < \infty, \tag{4.11}$$
$$\operatorname{Im}\ z > 0.$$

Theorem 4.1. *Suppose the system* (4.1) *has a unique Weyl-Titchmarsh function (is in the Weyl limit-point case) when* $\sigma = 0$*. Then for any* σ, $\sigma = \bar{\sigma}$, *the system* (4.1) *has a unique Weyl-Titchmarsh function, and the following equality holds:*

$$v(\sigma, z) = A(\sigma, z)\,[v(z)J]\,A^{-1}(\sigma, z)\,J. \tag{4.12}$$

Proof. Using formulas (4.3) and (4.7), we perform the shift $x+\sigma = u$ in inequalities (4.11):

$$\begin{aligned} \int_{\sigma}^{\infty} (J + iv^*(\sigma, z))A^{*-1}(\sigma, z)\mathcal{W}_+^*(u, z) \times \\ \times \mathcal{H}_+(u)\mathcal{W}_+(u, z)A^{-1}(\sigma, z)(J - iv(\sigma, z))\,du < \infty. \end{aligned} \tag{4.13}$$

Similarly, the shift $x - \sigma = u$ yields

$$\begin{aligned} \int_{-\sigma}^{\infty} (J - iv^*(\sigma, z))A^{*-1}(\sigma, z)\mathcal{W}_-^*(u, z) \times \\ \times \mathcal{H}_-(u)\mathcal{W}_-(u, z)A^{-1}(\sigma, z)(J + iv(\sigma, z))\,du < \infty. \end{aligned} \tag{4.14}$$

Since when $\sigma = 0$ we have a unique Weyl-Titchmarsh function (limit-point case), relations (4.13) and (4.14) yield

$$v(z) = A^{-1}(\sigma, z)v(\sigma, z)JA(\sigma, z)J,$$

which completes the proof of the theorem. □

Corollary 4.1. *The singular points* $z_k \neq \infty$ *of the matrix-valued functions* $v(z)$ *and* $v(\sigma, z)$ *coincide.*

The validity of this assertion follows directly from formula (4.12) if we take into account that $A(\sigma, z)$ and $A^{-1}(\sigma, z)$ are entire matrix-valued functions of z.

Chapter 10
A Class of Canonical Systems

This chapter is closely connected with M. G. Kreĭn's celebrated paper [48] in which a number of fundamental facts of the theory of differential systems were announced. First we shall recall the main notions and results of [48]

Consider a continuous function

$$H(t) = \overline{H(-t)}, \qquad -r \leq t \leq r, \tag{0.1}$$

We will assume that the operator S_r acting as

$$S_r f = f(x) + \int_0^r H(x-t)f(t)\,dt \tag{0.2}$$

is positive. Then there exists the Hermitian *resolvent kernel* $\Gamma_r(t,s) = \overline{\Gamma_r(s,t)}$, which satisfies the relation

$$\Gamma_r(t,s) + \int_0^r H(t-u)\Gamma_r(u,s)\,du = H(t-s), \qquad 0 \leq s,t \leq r. \tag{0.3}$$

Moreover, the following equalities hold (see [48]):

$$\frac{\partial \Gamma_r(t,s)}{\partial r} = -\Gamma_r(r,s)\Gamma_r(t,r), \tag{0.4}$$

$$\Gamma_r(t,s) = \Gamma_r(r-s, r-t). \tag{0.5}$$

Let us introduce the functions

$$P(r,\lambda) = e^{i\lambda r}(1 - \int_0^r \Gamma_r(s,0)e^{-i\lambda s}ds), \tag{0.6}$$

and

$$P_*(r,\lambda) = 1 - \int_0^r \Gamma_r(0,s)e^{i\lambda s}ds. \tag{0.7}$$

Then from (0.4) and (0.5) we obtain the system of ordinary differential equations

$$\begin{cases} \dfrac{dP(r,\lambda)}{dr} = i\lambda P(r,\lambda) - \overline{A(r)}P_*(r,\lambda), \\ \dfrac{dP_*(r,l)}{dr} = -A(r)P(r,\lambda), \end{cases} \tag{0.8}$$

where

$$A(r) = \Gamma_r(0,r). \tag{0.9}$$

The differential system (0.8) is now called the *Kreĭn system.* M. G. Kreĭn formulated the following important result.

Theorem 0.1 (Kreĭn [48]). *The following assertions are equivalent:*

1) *The integral*

$$K(z_0) = \int_0^\infty |P(r, z_0)|^2 \, dr \tag{0.10}$$

converges for at least one z_0 *with* $\operatorname{Im} z_0 > 0$.

2) *The function* $P_*(r, z_0)$ $(0 \leq x < \infty)$ *is bounded for at least one* z_0 *with* $\operatorname{Im} z_0 > 0$.

3) *The integral* $K(z)$ *converges uniformly on any bounded closed subset of the open half-plane* $\operatorname{Im} z > 0$.

4) *There exists the limit*

$$P(z) = \lim_{r \to \infty} P_*(r, z), \tag{0.11}$$

where convergence is uniform on any bounded closed subset of the open half-plane $\operatorname{Im} z > 0$.

In this chapter we study canonical systems of the form

$$\frac{d\mathcal{W}}{dx} = izJ\mathcal{H}(x)\mathcal{W}, \qquad J = \begin{bmatrix} 0 & I_m \\ I_m & 0 \end{bmatrix}, \qquad \mathcal{H}(x) \geq 0\,, \tag{0.12}$$

where the matrix $J\mathcal{H}(x)$ is assumed to be linearly similar to the matrix

$$M_1 = \begin{bmatrix} D & 0 \\ 0 & 0 \end{bmatrix}, \qquad D = \operatorname{diag}\{d_1, d_2, \dots, d_m\}, \qquad d_k > 0. \tag{0.13}$$

When $m = 1$ and $d_1 = 1$, the system (0.12) is equivalent to Kreĭn's system (0.8). When $m > 1$ and $d_j \neq d_k$, the system (0.12) generates non-classical spectral problems. Such non-classical problems play an essential role both in theoretical and applied domains.

10.1. Asymptotic formulas

Consider the canonical system (0.12). We will assume that

$$J\mathcal{H}(x) = T(x)M_1T^{-1}(x), \tag{1.1}$$

where

$$M_1 = \begin{bmatrix} D & 0 \\ 0 & 0 \end{bmatrix}, \qquad D = \operatorname{diag}\{d_1, d_2, \dots, d_m\}, \qquad d_k > 0.$$

With no loss of generality we can also assume that

$$T(x)jT^*(x) = J, \tag{1.2}$$

where

$$j = \begin{bmatrix} I_m & 0 \\ 0 & -I_m \end{bmatrix}.$$

Here we use a well-known theorem of Potapov ([65], Ch. 1, § 2) on the canonical form of J-Hermitian positive matrices. We shall further assume that the matrix-valued function $T(x)$ is continuously differentiable. Set

$$W(x,z) = T^{-1}(x)\mathcal{W}(x,z)T(0). \tag{1.3}$$

It follows from (1.1) that

$$\frac{dW}{dx} = izM_1W(x,z) + A(x)W(x,z), \tag{1.4}$$

where

$$A(x) = -T^{-1}(x)\,\frac{dT}{dx}. \tag{1.5}$$

Differentiating both sides of relation (1.2), we obtain

$$A(x)j = -jA^*(x). \tag{1.6}$$

Let us write the solution of system (1.4) satisfying the condition

$$W(0,z) = I_{2m} \tag{1.7}$$

in the following form:

$$W(x,z) = \exp\{izxM_1\}\overset{\curvearrowright}{\int_0^x} \exp\left[A(t,z)\,dt\right], \tag{1.8}$$

where

$$A(x,z) = \exp\{-izxM_1\}A(x)\exp\{izxM_1\}. \tag{1.9}$$

10.1.2. Let us write the matrices $W(x,z)$ and $A(x)$ in block form with blocks of order $m \times m$:

$$W(x,z) = \{W_{ij}(x,z)\}_{i,j=1}^2, \qquad A(x) = \{A_{ij}(x)\}_{i,j=1}^2. \tag{1.10}$$

By (1.6),

$$A_{11}(x) = -A_{11}^*(x), \qquad A_{12}^*(x) = A_{21}(x). \tag{1.11}$$

From relations (1.40) and (1.41) of Chapter 8 we obtain the following assertion.

Proposition 1.1. *The matrix $T(x)$ can be chosen in such a way that relations* (1.1) *and* (1.2) *are satisfied and*

$$A_{22}(x) = 0. \tag{1.12}$$

10.1.3. Consider the matrix-valued functions

$$S(x) = T^{-1}(x) = \begin{bmatrix} S_{11}(x) & S_{12}(x) \\ S_{21}(x) & S_{22}(x) \end{bmatrix}, \tag{1.13}$$

$$P_1(x,z) = W_{11}(x,z)S_{12}(0) + W_{12}(x,z)S_{22}(0), \tag{1.14}$$

$$P_2(x,z) = W_{21}(x,z)S_{12}(0) + W_{22}(x,z)S_{22}(0), \tag{1.15}$$

and

$$P(x,z,\zeta) = P_2^*(x,z)P_2(x,\zeta) - P_1^*(x,z)P_1(x,\zeta). \tag{1.16}$$

By (1.4), the following equations hold:

$$\begin{cases} \dfrac{dP_1}{dx} = izDP_1 + A_{11}(x)P_1 + A_{12}(x)P_2, \\ \dfrac{dP_2}{dx} = A_{21}(x)P_1. \end{cases} \tag{1.17}$$

Differentiating both sides of relation (1.16) with respect to x and taking into account relations (1.11), (1.12), and (1.17), we obtain the equation

$$\frac{dP}{dx} = i(\bar{z} - \zeta)P_1^*(x,z)DP_1(x,\zeta). \tag{1.18}$$

Next, by (1.2), we have

$$S^*(x)jS(x) = J, \tag{1.19}$$

whence, in particular,

$$S_{12}^*(0)\,S_{12}(0) = S_{22}^*(0)\,S_{22}(0). \tag{1.20}$$

In conjunction with formulas (1.14)–(1.16), this implies that

$$P_1(0,z) = S_{12}(0), \qquad P_2(0,z) = S_{22}(0), \tag{1.21}$$

$$P(0,z,\zeta) = 0. \tag{1.22}$$

Equalities (1.18) and (1.22) are equivalent to the relation

$$P_2^*(x,z)P_2(x,\zeta) - P_1^*(x,z)P_1(x,\zeta) = i(\bar{z} - \zeta)\int_0^x P_1^*(t,z)DP_1(t,\zeta)\,dt. \tag{1.23}$$

Setting here $\zeta = z$, we get

$$P_2^*(x,z)P_2(x,z) - P_1^*(x,z)P_1(x,z) = 2\,\mathrm{Im}\,z\int_0^x P_1^*(t,z)DP_1(t,z)\,dt. \tag{1.24}$$

Further, from relation (1.24) it follows that

$$\det P_1(x,z) \neq 0, \qquad \mathrm{Im}\,z \leq 0, \tag{1.25}$$

and

$$\det P_2(x,z) \neq 0, \qquad \mathrm{Im}\,z \geq 0. \tag{1.26}$$

Relation (1.24) plays an essential role in proving the following assertion.

Theorem 1.1 (Generalized Kreĭn Theorem). *The following assertions are equivalent*

1) *The integral*

$$K(z_0) = \int_0^\infty P_1^*(x, z_0) D P_1(x, z_0)\, dx \tag{1.27}$$

converges for at least one z_0 *with* $\operatorname{Im} z_0 > 0$.

2) *The norm of the matrix-valued function* $P_2(x, z_0)$ $(0 \leq x < \infty)$ *is bounded for at least one* z_0 *with* $Im\, z_0 > 0$.

3) *The integral* $K(z)$ *converges uniformly on any bounded closed subset of the open half-plane* $\operatorname{Im} z > 0$.

4) *There exists the limit*

$$\mathcal{P}(z) = \lim_{x \to \infty} P_2(x, z), \tag{1.28}$$

where convergence is uniform on any bounded closed subset of the open half-plane $\operatorname{Im} z > 0$.

Proof. Consider the integral

$$K_R(z, \zeta) = \int_0^R P_1^*(x, z) D P_1(x, \zeta)\, dx. \tag{1.29}$$

Relation (1.24) yields

$$P_2^*(x, z_0) P_2(x, z_0) \geq 2 \operatorname{Im} z_0\, K_R(z_0, z_0), \quad x \geq R. \tag{1.30}$$

Now set

$$\varphi_R(z_0) = g^* K_R(z_0, z_0) g,$$

where g is a column-vector, $g \neq 0$.

The function $\varphi_R(z_0)$ increases monotonically with R and has either a finite or infinite limit

$$\lim_{R \to \infty} \varphi_R(z_0) = \varphi(z_0) \leq \infty. \tag{1.31}$$

Let Q be an arbitrary closed bounded subset of the open half-plane $\operatorname{Im}\ z > 0$. Thanks to inequality (1.30), for any choice of numbers R_n, $R_n \to \infty$ as $n \to \infty$, the sequence $\{g^* P_2(R_n, z) g\}$ is uniformly bounded from below in Q. By Montel's theorem, there exists a subsequence which converges uniformly to a limit

$$\mathcal{P}(z, g) = \lim_{R_n \to \infty} g^* P_2(R_n, z) g, \qquad z \in Q. \tag{1.32}$$

Moreover, here

$$\mathcal{P}(z, g) \neq 0, \qquad \operatorname{Im} z > 0. \tag{1.33}$$

We claim that the conditions of the theorem exclude the case

$$\mathcal{P}(z,g)=\infty. \tag{1.34}$$

Indeed, let condition 1) be satisfied. Then there exists a sequence $\{P_1(R_k,z_0)\}$ such that

$$P_1(R_k,z_0)\to 0. \tag{1.35}$$

In conjunction with (1.24), this shows that the sequence $P_2(R_k,z_0)$ is bounded, i.e., $\mathcal{P}(z_0,g)\neq\infty$. Hence, conditions 2) and 4) are satisfied. Thanks to (1.24), condition 3) is also satisfied. Thus we have proved that conditions 2)–4) follow from condition 1). The remaining equivalences of the theorem can be proved in a similar fashion. □

Suppose now that one of the conditions of Theorem 1.1 is satisfied. Then from (1.24) and (1.29) we deduce that

$$\lim_{x\to\infty} P_1(x,z)=0. \tag{1.36}$$

Further, relations (1.23) and (1.36) yield the equality

$$\mathcal{P}^*(z)\mathcal{P}(\zeta)=i(\bar{z}-\zeta)\int_0^\infty P_1^*(x,z)DP_1(x,\zeta)\,dx \tag{1.37}$$

for $\operatorname{Im} z>0$, $\operatorname{Im}\zeta>0$.

Let us examine separately the case when

$$\|A(\cdot)\|\in L(0,\infty). \tag{1.38}$$

Lemma 1.1. *Under assumption* (1.38), *the inequality*

$$\|W(x,z)\|\le C,\qquad 0\le x<\infty,\qquad \operatorname{Im} z\ge 0. \tag{1.39}$$

holds for some constant $C>0$.

Proof. Setting $x_k=x\frac{k}{N}$ $(0\le k\le N)$, one has that

$$W(x,z)=\lim_{N\to\infty}\left[U(x_N,x_{N-1})U(x_{N-1},x_{N-2})\cdot\ldots\cdot U(x_1,x_0)\right], \tag{1.40}$$

where

$$U(x_k,x_{k-1})=\exp\left[\int_{x_{k-1}}^{x_k} A(x)\,dx\right]\exp\left[iz(x_k-x_{k-1})M_1\right]. \tag{1.41}$$

Consequently,

$$\|W(x,z)\|\le\exp\left[\int_0^x \|A(x)\|\,dx\right], \tag{1.42}$$

which thanks to (1.38) completes the proof of the lemma. □

Corollary 1.1. *If condition* (1.38) *is satisfied, then all the assertions of Theorem* 1.1 *hold true.*

Indeed, by the estimate (1.39), assertion 2) is true, and then so are the other equivalent assertions.

Corollary 1.2. *Let condition* (1.38) *be satisfied. Then the following assertions are valid:*

1) *In the closed upper plane* $\operatorname{Im} z \geq 0$ *there exists the limit*

$$W_+(z) = \lim_{x\to\infty} W(x,z),$$

and the matrix-valued function $W_+(z)$ *is continuous.*

2) *The matrix-valued function* $\mathcal{P}(z)$ *is continuous in the closed upper half-plane* $\operatorname{Im} z \geq 0$.

Lemma 1.2. *Suppose the condition*

$$\int_0^\infty \|A_{21}\|\,dx < \infty \tag{1.43}$$

is satisfied. Then the following estimate holds:

$$\|\mathcal{P}^{\pm 1}(z)\| \leq \|S_{22}^{\pm 1}(0)\| \exp\left(\int_0^\infty \|A_{21}(x)\|\,dx\right). \tag{1.44}$$

Proof. Integrating equation (1.7), we have

$$P_2(x,z) = S_{22}(0) + \int_0^x A_{21}(t)P_1(t,z)\,dt. \tag{1.45}$$

This relation can be recast as

$$P_2(x,z) = S_{22}(0) + \int_0^x A_{21}\left[P_1(x,z)P_2^{-1}(x,z)\right]\,dx. \tag{1.46}$$

Since, thanks to (1.24),

$$\|P_1(x,z)P_2^{-1}(x,z)\| \leq 1, \qquad \operatorname{Im} z \geq 0 \tag{1.47}$$

representation (1.46) yields

$$\|P_2(x,z)\| \leq \|S_{22}(0)\| + \int_0^x \|A_{21}(x)\| \cdot \|P_2(x,z)\|\,dx. \tag{1.48}$$

As is well known (see [10]), this in turn yields the estimate

$$\|P_2(x,z)\| \le \|S_{22}(0)\| \; \exp\left(\int_0^x \|A_{21}(x)\|\,dx\right). \tag{1.49}$$

Letting here $x \to \infty$, we obtain

$$\|\mathcal{P}(z)\| \le \|S_{22}(0)\| \; \exp\left(\int_0^\infty \|A_{21}(x)\|\,dx\right). \tag{1.50}$$

In order to obtain an estimate of $\mathcal{P}^{-1}(z)$ let us consider the derivative

$$\begin{aligned} \frac{d}{dx} P_2^{-1}(x,z) &= -P_2^{-1}(x,z)\left[\frac{d}{dx} P_2(x,z)\right] P_2^{-1}(x,z) = \\ &= -P_2^{-1}(x,z) A_{21}(x) P_1(x,z) P_2^{-1}(x,z). \end{aligned} \tag{1.51}$$

Integrating both sides of this equality, we obtain

$$P_2^{-1}(x,z) = S_{22}^{-1}(0) - \int_0^x P_2^{-1}(t,z) A_{21}(t) P_1(t,z) P_2^{-1}(t,z)\,dt.$$

Using inequality (1.47) one more time, we have

$$\|P_2^{-1}(x,z)\| \le \|S_{22}^{-1}(0)\| + \int_0^x \|P_2^{-1}(t,z)\| \cdot \|A_{21}(t)\|\,dt,$$

i.e.,

$$\|P_2^{-1}(x,z)\| \le \|S_{22}^{-1}(0)\| \;\; \exp\left(\int_0^x \|A_{21}(t)\|\,dt\right), \tag{1.52}$$

Letting here $x \to \infty$, we obtain

$$\|\mathcal{P}^{-1}(z)\| \le \|S_{22}^{-1}(0)\| \exp\left(\int_0^x \|A_{21}(x)\|\,dx\right),$$

which completes the proof of the lemma. □

Let us remark that, thanks to relations (1.19) and (1.20),

$$\det S_{22}(0) \ne 0,$$

and so the matrix $S_{22}(0)$ is invertible.

Letting $x \to \infty$ in (1.45) we obtain the relation

$$\mathcal{P}(z) = S_{22}(0) + \int_0^\infty A_{21}(t) P_1(t,z)\,dt. \tag{1.53}$$

Remark 1.1. It follows from Lemma 1.2 that all the assertions of Theorem 1.1 hold true whenever condition (1.43) is fulfilled. This condition is somewhat weaker than (1.38).

Remark 1.2. For $m = 1$ the system (1.4) was introduced and investigated by M. G. Kreĭn in his celebrated article [48]. Unfortunately, he did not publish complete proofs of the results formulated therein.

In this section we generalized a number of results of M. G. Kreĭn [48] for the case when $m \geq 1$ and give the corresponding proofs.

Remark 1.3. The assertions of Theorem 1.1 and of Lemmas 1.1, 1.2 are continual matrix analogues of results obtained for orthogonal polynomials [97].

10.2. Spectral analysis

10.2.1. In this section we shall apply the general spectral theorems obtained in Chapter 4 to the case when conditions (1.1) and (1.2) are fulfilled. Set

$$\beta(x) = [\sqrt{D}\ \ 0]T^*(x)J, \tag{2.1}$$

$$\mathcal{H}(x) = \beta^*(x)\beta(x). \tag{2.2}$$

Consider the operator A_{M} acting in the space $L^2_m(0,l)$ associated with system (0.12) by the rule (see [45])

$$A_{\mathrm{M}}f = i\beta(x)J\int_0^x \beta^*(t)f(t)\,dt. \tag{2.3}$$

We shall use the decomposition of the matrix $T(x)$ into $m \times m$ blocks:

$$T(x) = \begin{bmatrix} t_{11}(x) & t_{12}(x) \\ t_{21}(x) & t_{22}(x) \end{bmatrix}.$$

It follows from (2.1) that

$$\beta(x) = [\beta_1(x)\ \ \beta_2(x)], \tag{2.4}$$

where the matrices $\beta_1(x)$ and $\beta_2(x)$ are defined as

$$\beta_1(x) = \sqrt{D}\,t^*_{21}(x), \qquad \beta_2(x) = \sqrt{D}\,t^*_{11}(x). \tag{2.5}$$

Definition 2.1. A pair of meromorphic matrix-valued functions $R(z)$, $Q(z)$ ($\mathrm{Im} z > 0$) is called a *non-special pair with the J-property* if

$$\det[R^*(z)R(z) + Q^*(z)Q(z)] \not\equiv 0, \qquad R^*(z)Q(z) + Q^*(z)R(z) \geq 0, \qquad \mathrm{Im}\ z > 0.$$

Theorem 2.1. *Let conditions* (1.1) *and* (1.2) *be satisfied and let the matrix-valued function* $T(x)$ *be continuously differentiable.*

Then the set of the Weyl-Titchmarsh matrix functions $v(z)$ *of the corresponding system* (0.12) *is described by means of the formula*

$$v(z) = i\,[a(z)R(z) + b(z)Q(z)]\,[c(z)R(z) + d(z)Q(z)]^{-1}, \qquad \operatorname{Im} z > 0 \tag{2.6}$$

where $R(z)$, $Q(z)$ *is an arbitrary non-special pair with the* J*-property and the matrix of coefficients has the form*

$$\mathfrak{A}(l,z) = \begin{bmatrix} a(z) & b(z) \\ c(z) & d(z) \end{bmatrix} = \mathcal{W}^*(l,\bar{z}). \tag{2.7}$$

Proof. Let us show that $\lambda = 0$ is not an eigenvalue of the operator A_{M}. Assume the contrary, i.e., there exists a vector-valued function f, $\|f_0\| \neq 0$, such that

$$A_{\mathrm{M}} f_0 = 0. \tag{2.8}$$

Differentiating both sides of (2.8), we obtain

$$iDf_0(x) + \int_0^x K(x,t) f_0(t)\,dt = 0, \tag{2.9}$$

where

$$K(x,t) = \sqrt{D}\,[t_{11}'(x)^* t_{21}(t) + t_{21}'(x)^* t_{11}(t)]\,\sqrt{D}. \tag{2.10}$$

Here we used the fact that

$$T^*(x) J T(x) = j. \tag{2.11}$$

However, equality (2.9) is possible only when $\|f_0\| = 0$. We reached a contradiction, and so $\lambda = 0$ is not an eigenvalue of A_{M}. From (1.2) it follows that

$$t_{11}(x)t_{11}^*(x) - t_{12}(x)t_{12}^*(x) = 0 \tag{2.12}$$

and

$$t_{11}(x)t_{21}^*(x) - t_{12}(x)t_{22}^*(x) = I_m. \tag{2.13}$$

Finally, (2.12) implies that for any nonzero vector $h = [h_1, h_2, \ldots, h_m]$ one has simultaneously that $ht_{11} = 0$ and $ht_{12} = 0$. By (2.13), this imples that $h = 0$. Thus,

$$\det t_{11}(x) \neq 0, \tag{2.14}$$

and consequently all the conditions of the general Theorem 3.3 of Chapter 4 are satisfied. This completes the proof of our theorem. □

10.2.2. The spectral data $\tau(u)$ and α of the system (0.12) are connected with the Weyl-Titchmarsh matrix function of that system via the equality (see Chapter 4)

$$v(z) = \alpha + \int_{-\infty}^{\infty} \left(\frac{1}{u-z} - \frac{u}{1+u^2} \right) d\tau(u). \tag{2.15}$$

According to the general theory (see Chapter 4), the appropriate analogue of the Fourier transformation has the form

$$F(u) = Vg(x) = \int_0^{\infty} [0 \ I_m] \mathcal{W}^*(x,u)\mathcal{H}(x)g(x)\,dx, \tag{2.16}$$

and the Parseval relation holds:

$$\int_{-\infty}^{\infty} F^*(u)[d\tau(u)]F(u) = \int_0^{\infty} g^*(x)\mathcal{H}(x)g(x)\,dx. \tag{2.17}$$

Let us rewrite relations (2.16) and (2.17) in terms of the system (1.4). By (1.3),

$$Vg = \int_0^{\infty} [0 \ I_m] T^{*-1}(0)W^*(x,u)T^*(x)\mathcal{H}(x)g(x)\,dx. \tag{2.18}$$

Using relations (2.1), (2.2) we reduce (2.18) to the form

$$F(u) = Vg = \int_0^{\infty} [0 \ I_m] T^{*-1}(0)\, W^*(x,u)\,dx. \tag{2.18}$$

Using relations (2.1) and (2.2) one can recast (2.18) as

$$F(u) = Vg = \int_0^{\infty} [0 \ I_m] T^{*-1}(0)W^*(x,u) \begin{bmatrix} D & 0 \\ 0 & 0 \end{bmatrix} T^{-1}(x)\, g(x)\,dx. \tag{2.19}$$

If we write the vector $g(x)$ in the block form $g(x) = \text{col}\,[g_1(x), g_2(x)]$, where $g_k(x)$ is of order $m \times 1$, then (2.19) yields

$$F(u) = \int_0^{\infty} P_1^*(x,u)Df(x)\,dx, \tag{2.20}$$

where $P_1(x,u)$ is defined by formula (1.14) and

$$f(x) = S_{11}(x)g_1(x) + S_{12}(x)g_2(x). \tag{2.21}$$

Using relations (2.1), (2.2), and (2.21), we can rewrite equality (2.17) in the form

$$\int_{-\infty}^{\infty} F^*(u)[d\tau(u)]F(u) = \int_0^{\infty} f^*(x)Df(x)\,dx. \tag{2.22}$$

Thus the spectral functions of system (1.4) are defined by relations (2.20), (2.22).

10.2.3. Now let us take the system (1.4), where $A(x)$ satisfies conditions (1.11) and (1.12), as the starting point of our investigation. The matrix $T(x)$ is defined in this case by relation (1.5), and $T(0)$ satisfies the condition

$$T^*(0)JT(0) = j. \tag{2.23}$$

Remark 2.1. Apart from condition (2.32), the matrix $T(0)$ will be assumed to be arbitrary.

The matrix-valued functions $P_1(x,z)$ and $P_2(x,z)$ are solutions of the system (1.4) and satisfy the boundary conditions (1.21). Hence, (2.20) corresponds to the boundary problem (1.4), (1.21).

Let us examine separately the special case when

$$T(0) = \frac{1}{\sqrt{2}} \begin{bmatrix} I_m & -I_m \\ I_m & I_m \end{bmatrix}. \tag{2.24}$$

Under this assumption,

$$S(0) = \frac{1}{\sqrt{2}} \begin{bmatrix} I_m & I_m \\ -I_m & I_m \end{bmatrix}. \tag{2.25}$$

which in conjunction with (1.14) and (1.15) implies that in the present case

$$P_1(0,z) = P_2(0,z). \tag{2.26}$$

Remark 2.2. If formulas (1.2) and (1.3) are used, then equality (2.6) can be recast it the terms of problem (1.4), (1.21):

$$v(z) = i\left[a_1(z)R_1(z) + b_1(z)Q_1(z)\right]\left[c_1(z)R_1(z) + d_1(z)Q_1(z)\right]^{-1}, \tag{2.27}$$

where the matrix of coefficients $\mathfrak{A}_1(z)$ has the form

$$\mathfrak{A}_1(z) = \begin{bmatrix} a_1(z) & b_1(z) \\ c_1(z) & d_1(z) \end{bmatrix} = T^{*-1}(0)W^*(l,\bar{z}), \tag{2.28}$$

and where $R_1(z)$, $Q_1(z)$ is a non-special pair with the j-property , i.e.,

$$\det\left[R_1^*(z)R_1(z) + Q_1^*(z)Q_1(z)\right] \not\equiv 0, \tag{2.29}$$

$$R_1^*(z)R_1(z) - Q_1^*(z)Q_1(z) \geq 0, \qquad \operatorname{Im} z > 0. \tag{2.30}$$

Remark 2.3. For the boundary value problem (1.4), (2.26) with $m = 1$ and $D = 1$, formula (2.20) and the Parseval identity (2.22) were derived by M. G. Kreĭn in [48].

10.3. Transformed canonical systems

10.3.1. With the help of a simple substitution we can transform canonical systems from one class to another. Indeed, let us set

$$\widetilde{\mathcal{W}}(x,z) = e^{-izx\gamma}\mathcal{W}(x,z), \qquad \gamma > 0. \tag{3.1}$$

Then system (0.12) turns into the system

$$\frac{d\widetilde{\mathcal{W}}}{dx} = izJ\widetilde{\mathcal{H}}(x)\widetilde{\mathcal{W}}(x,z), \tag{3.2}$$

where

$$J\widetilde{\mathcal{H}}(x) = J\mathcal{H}(x) - \gamma I_{2m}. \tag{3.3}$$

The new system (3.2) is canonical if and only if

$$\widetilde{\mathcal{H}}(x) \geq 0. \tag{3.4}$$

In the case of system (0.12), (0.13), condition (3.4) is equivalent to the inequality

$$D - \gamma I_m \geq 0. \tag{3.5}$$

The following assertions hold true.

Proposition 3.1. *The set of the Weyl-Titchmarsh matrix functions $v(z)$ of system (0.12) on the interval $[0,l]$ coincides with the set of the Weyl-Titchmarsh matrix functions of the system (3.2) on the interval $[0,l]$.*

Proof. By the definition of the transformation (3.1), the matrix

$$\widetilde{\mathcal{W}}(l,\bar{z}) = \begin{bmatrix} \widetilde{a}(l,z) & \widetilde{b}(l,z) \\ \widetilde{c}(l,z) & \widetilde{d}(l,z) \end{bmatrix}$$

differs from the matrix

$$\mathcal{W}(l,\bar{z}) = \begin{bmatrix} a(l,z) & b(l,z) \\ c(l,z) & d(l,z) \end{bmatrix}$$

only by a scalar factor. Consequently,

$$\begin{aligned}&[a(l,z)R(z) + b(l,z)Q(z)]\ [c(l,z)R(z) + d(l,z)Q(z)]^{-1} \\ &= \left[\widetilde{a}(l,z)R(z) + \widetilde{b}(l,z)Q(z)\right]\ \left[\widetilde{c}(l,z)R(z) + \widetilde{d}(l,z)Q(z)\right]^{-1}.\end{aligned}$$

To complete the proof it remains to use formula (2.26) of Chapter 4. □

Next, Proposition 3.1 and the Stieltjes inversion formula yield the following result.

Proposition 3.2. *The spectral data of systems* (0.12) *and* (3.2) *coincide.*

Finally, using Theorem 2.2 of Chapter 9 one derives the following assertion.

Proposition 3.3. *Let conditions* (1.1) *and* (1.2), *where* $T(x)$ *is a continuous matrix-valued function, be satisfied. Then the system* (0.12) *on the half-line* $0 \leq x < \infty$ *is in the Weyl limit-point case.*

Remark 3.1. Propositions 3.1 and 3.2 show that fixed spectral data $\tau(u)$ and α can generate different canonical systems, i.e., the corresponding inverse spectral problem has more than one solution.

10.3.2. Set

$$\widetilde{W}(x,z) = T^{-1}(x)\widetilde{\mathcal{W}(x,z)}T(0). \tag{3.6}$$

Then from (3.2) one derives the relation

$$\frac{d\widetilde{W}(x,z)}{dx} = izM_2\widetilde{W}(x,z) + A(x)\widetilde{W}(x,z), \tag{3.7}$$

where

$$M_2 = M_1 - I_{2m}\gamma = \begin{bmatrix} D - \gamma I_{2m} & 0 \\ 0 & -\gamma I_m \end{bmatrix}. \tag{3.8}$$

Now let us introduce the matrices

$$\widetilde{P}_1(x,z) = \widetilde{W}_{11}(x,z)S_{12}(0) + \widetilde{W}_{12}(x,z)S_{22}(0), \tag{3.9}$$

$$\widetilde{P}_2(x,z) = \widetilde{W}_{21}(x,z)S_{12}(0) + \widetilde{W}_{22}(x,z)S_{22}(0). \tag{3.10}$$

Then, by (3.7),

$$\begin{cases} \dfrac{d\widetilde{P}_1}{dx} = iz(D - \gamma I_m)\widetilde{P}_1 + A_{11}(x)\widetilde{P}_1 + A_{12}(x)\widetilde{P}_2, \\ \dfrac{d\widetilde{P}_2}{dx} = -iz\gamma\widetilde{P}_2 + A_{21}(x)\widetilde{P}_1. \end{cases} \tag{3.11}$$

Setting $x = 0$ in relations (3.9) and (3.10) yields

$$\widetilde{P}_1(0,z) = S_{12}(0), \qquad \widetilde{P}_2(0,z) = S_{22}(0). \tag{3.12}$$

The boundary value problem (3.11), (3.12) is of great interest in its own right. In this case the general formula (2.18) has the form

$$F(u) = Vf = \int_0^\infty \left[\widetilde{P}_1^*(x,u)\ \widetilde{P}_2^*(x,u)\right] D_1 f(x)\,dx, \tag{3.13}$$

where

$$D_1 = \begin{bmatrix} D - \gamma I_m & 0 \\ 0 & \gamma I_m \end{bmatrix}, \qquad f(x) = \text{col}\,\left[f_1(x), \dots, f_{2m}(x)\right].$$

The corresponding Parseval equality reads

$$\int_{-\infty}^{\infty} F^*(u)[d\tau(u)]F(u) = \int_0^{\infty} f^*(x)D_1 f(x)\,dx. \tag{3.14}$$

Thus, the spectral functions of the system (3.11), (3.12) are defined by relations (3.13) and (3.14). The description of the set of the Weyl-Titchmarsh functions $v(z)$ of the system (3.11), (3.12) on the interval $[0,l]$ is given by the formula

$$v(z) = i[\widetilde{a}_1(z)R_1(z) + \widetilde{b}_1(z)Q_1(z)]\,[\widetilde{c}_1(z)R_1(z) + \widetilde{d}_1(z)Q_1(z)]^{-1}, \tag{3.15}$$

where the matrix of coefficients $\widetilde{\mathfrak{A}}_1(z)$ has the form

$$\widetilde{\mathfrak{A}}_1(z) = \begin{bmatrix} \widetilde{a}_1(z) & \widetilde{b}_1(z) \\ \widetilde{c}_1(z) & \widetilde{d}_1(z) \end{bmatrix} = T^{*-1}(0)\widetilde{W}^*(l,\bar{z}), \tag{3.16}$$

and where the pairs $R_1(z)$, $Q_1(z)$ satisfy the conditions (2.29), (2.30). Formulas (3.13)–(3.16) are deduced in much the same fashion as formulas (2.20), (2.22) and (2.27), (2.28).

10.4. Dirac-type systems

10.4.1. Let us consider separately the important special case of systems (3.11) when

$$D = I_m, \qquad A_{11}(x) = 0, \qquad \gamma = \frac{1}{2}. \tag{4.1}$$

In this case the system (3.11) has the form

$$\begin{cases} \dfrac{d\widetilde{P}_1}{dx} = i\dfrac{z}{2}\widetilde{P}_1 + A_{12}(x)\widetilde{P}_2, \\ \dfrac{d\widetilde{P}_2}{dx} = -i\dfrac{z}{2}\widetilde{P}_2 + A_{12}^*(x)\widetilde{P}_1. \end{cases} \tag{4.2}$$

Let us introduce the following notations:

$$\Phi(x,z) = \frac{1}{2}[\widetilde{P}_1(x,z) + \widetilde{P}_2(x,z)], \tag{4.3}$$

$$\Psi(x,z) = \frac{i}{2}[\widetilde{P}_1(x,z) - \widetilde{P}_2(x,z)], \tag{4.4}$$

$$a(x) = \frac{1}{2}[A_{12}(x) + A_{12}^*(x)], \qquad b(x) = \frac{1}{2i}[A_{12}(x) - A_{12}^*(x)]. \tag{4.5}$$

Then the system (4.2) can be rewritten as

$$\begin{cases} \dfrac{d\Phi}{dx} = & \dfrac{z}{2}\Psi + a\Phi - b\Psi, \\ \dfrac{d\Psi}{dx} = & -\dfrac{z}{2}\Phi - b\Phi - a\Psi. \end{cases} \tag{4.6}$$

Systems of the form (4.6) are called *Dirac-type systems.* By (3.12), (4.3), and (4.4), we have

$$\Phi(0,z) = \frac{1}{2}[S_{12}(0) + S_{22}(0)], \qquad \Psi(0,z) = \frac{i}{2}[S_{12}(0) - S_{22}(0)]. \tag{4.7}$$

If $S(0)$ has the form (2.25), then the boundary conditions (4.7) become

$$\Phi(0,z) = I_m, \qquad \Psi(0,z) = 0. \tag{4.8}$$

Also, for the Dirac-type system (4.6) relations (3.13) and (3.14) read

$$F(u) = Vg(x) = \int_0^\infty [\Phi^*(x,u) \;\; \Psi^*(x,u)]\, g(x)\, dx, \tag{4.9}$$

and

$$\int_{-\infty}^{\infty} F^*(u)[d\tau(u)]F(u) = \int_0^\infty g^*(x)g(x)\, dx, \tag{4.10}$$

respectively. Here we have used the fact that

$$\left[\widetilde{P}_1^*(x,u) \;\; \widetilde{P}_2^*(x,u)\right] = \left[\widetilde{\Phi}^*(x,u) \;\; \widetilde{\Psi}^*(x,u)\right] \begin{bmatrix} I_m & I_m \\ iI_m & -iI_m \end{bmatrix}. \tag{4.11}$$

We will assume in addition that

$$b(x) = 0 \tag{4.12}$$

and that the matrix-valued function $a(x)$ is differentiable.

Under these assumptions the first equation of system (4.6) yields

$$\Psi(x,z) = \frac{2}{z}\left(\frac{d\Phi}{dx} - a\Phi\right). \tag{4.13}$$

Using (4.13) and the second equation of system (4.6), we obtain

$$\frac{d^2\Phi}{dx^2} - \left[a'(x) + a^2(x)\right]\Phi + \frac{z^2}{4}\Phi = 0, \tag{4.14}$$

$$\Phi(0,z) = \frac{1}{2}[S_{12}(0) + S_{22}(0)], \tag{4.15}$$

$$\Phi'(0,z) - a(0)\Phi(0,z) = \frac{z}{2}\Psi(0,z). \tag{4.16}$$

Similarly, we deduce the relations

$$\frac{d^2\Psi}{dx^2} + [a'(x) - a^2(x)]\Psi + \frac{z^2}{4}\Psi = 0, \tag{4.17}$$

$$\Psi(0) = \frac{i}{2}[S_{12}(0) - S_{22}(0)], \tag{4.18}$$

$$\Psi'(0,z) + a(0)\Psi(0,z) = -\frac{z}{2}\Phi(0,z). \tag{4.19}$$

Thus in the case $b(x) = 0$ system (4.6) splits into two equations of second order. If $S(0)$ has form (2.25), then the boundary conditions split as well:

$$\Phi(0,z) = \frac{1}{\sqrt{2}} I_m, \qquad \Phi'(0,z) - a(0)\Phi(0,z) = 0, \tag{4.20}$$

$$\Psi(0,z) = 0, \qquad \Psi'(0,z) = -\frac{z}{2}. \tag{4.21}$$

10.5. An inverse problem

Consider the operator A acting as

$$Af = iD\int_0^x f(t)\,dt, \qquad 0 \le x \le l, \tag{5.1}$$

where

$$f = \mathrm{col}[f_1, f_2, \dots, f_m], \qquad D = \mathrm{diag}\,\{d_1, d_2, \dots, d_m\}, \quad d_k = \bar{d}_k \ne 0.$$

Define the operators Φ_2 and P_ζ by the rules

$$\Phi_2 g = g, \qquad P_\zeta f = \begin{cases} & \text{for } 0 \le x \le \zeta \\ 0 & \text{for } \zeta < x; \end{cases} \tag{5.2}$$

here g are constant $m \times 1$ vectors.

We begin our discussion with the simplest case in which

$$S_0 = I, \qquad \Phi_1 = \frac{1}{2}D, \qquad \alpha = 0. \tag{5.3}$$

In this case we have

$$J\mathcal{H}_0 = \frac{d}{dx}(J\Pi^* P_x \Pi) = \begin{bmatrix} \frac{1}{2}D & I_m \\ \frac{1}{4}D^2 & \frac{1}{2}D \end{bmatrix}.$$

It follows from (5.1) and (5.2) that

$$(I - A\lambda)^{-1}\Phi_2 = \exp\{i\lambda x D\}\Phi_2. \tag{5.5}$$

Hence, the operator $S_0 = E$ admits the representation

$$S_0 = \int_{-\infty}^{\infty} (I - A\lambda)^{-1}\Phi_2[d\tau_0(\lambda)]\Phi_2^*(I - A^*\lambda)^{-1}, \tag{5.6}$$

where

$$\tau_0(\lambda) = \frac{1}{2\pi}\lambda|D|, \qquad |D| = \text{diag }\{|d_1|, |d_2|, \dots, |d_m|\}. \tag{5.7}$$

From representation (5.6) we deduce that $\tau_0(\lambda)$ is a spectral matrix function of the system

$$\frac{dY}{dx} = izJ\mathcal{H}_0 Y, \qquad 0 \le x < \infty.$$

Remark 5.1. $\mathcal{H}_0(x) \in \mathbf{N}(M_1)$, i.e. the matrix JH_0 is linearly similar to the matrix

$$M_1 = \begin{bmatrix} D & 0 \\ 0 & 0 \end{bmatrix}.$$

Applying the general theory (Chapter 4) to the particular case (5.1), (5.2) we obtain the following result.

Theorem 5.1. *Let the operators A, Φ_2 and P_ζ be defined by formulas* (5.1), (5.2). *Assume that the monotonically increasing matrix-valued function $\tau(\lambda)$ satisfies the following conditions:*

1) *The matrix-valued function*

$$K(x,u) = \int_{-\infty}^{\infty} \exp\{i\lambda x D\}\{d[\tau(\lambda) - \tau_0(\lambda)]\}\exp\{-i\lambda x D\} \tag{5.9}$$

is continuous in the domain $0 \le x, u \le l$.

2) *For any vector-valued function $f(x)$, denote*

$$F(\lambda) = \int_{-\infty}^{\infty} \exp\{-i\lambda x D\} f(x)\, dx; \tag{5.10}$$

then

$$\int_{-\infty}^{\infty} F^*(\lambda)[d\tau(\lambda)]F(\lambda) = 0$$

implies $F(\lambda) = 0$.

Then $\tau(\lambda)$ is a spectral matrix function of the system (0.12). Moreover, the corresponding Hamiltonian $\mathcal{H}(x)$ belongs to the class $\mathbf{N}(M_1)$ and can be found by means of the chain of formulas

$$S_\zeta f = f(x) + \int_0^\zeta K(x,u)f(u)\,du, \tag{5.11}$$

$$\Pi = [\Phi_1 \ \ \Phi_2], \qquad \Phi_1 = -i\int_{-\infty}^{\infty} \frac{1}{\lambda(1+\lambda^2)}(\exp\{i\lambda xD\} - I_m)d\tau(\lambda) + i\Phi_2\alpha, \tag{5.12}$$

$$\mathcal{H}(\zeta) = \frac{d}{d\zeta}(\Pi^* S_\zeta^{-1} P_\zeta \Pi), \tag{5.13}$$

where $\alpha = \alpha^*$ is an $m \times m$ matrix.

Proof. The general formula for the operator S has the form (see Chapter 4)

$$S = \int_{-\infty}^{\infty} (I - A\lambda)^{-1}\Phi_2[d\tau(\lambda)]\Phi_2^*(I - A^*\lambda)^{-1}. \tag{5.14}$$

Since in the case under consideration (5.5) holds, equation (5.11), where $S_\zeta = P_\zeta S P_\zeta$, is seen to be a consequence of relations (5.2) and (5.9). The general formula for the operator Φ_1 has the form (see Chapter 4)

$$\Phi_1 = -i\int_{-\infty}^{\infty} \left[A(I - A\lambda)^{-1} + \frac{\lambda}{1+\lambda^2} I\right] \Phi_2 d\tau(\lambda) + i\Phi_2\alpha. \tag{5.15}$$

In the case (5.1), (5.2), formula (5.15) coincides with (5.12). It follows from representation (5.14) that $S \geq 0$. Thanks to assumptions 1), 2) of the theorem, the strict inequality

$$S \geq \delta I, \qquad \delta > 0 \tag{5.16}$$

holds. The existence and boundedness of the operator S_ζ^{-1} follows from (5.11) and (5.16). Hence, formula (5.13) is valid. Finally, by Corollary 2.1 (Chapter 7), the Hamiltonian $\mathcal{H}(x)$ belongs to the class $\mathbf{N}(M_1)$. The theorem is proved. □

Remark 5.2. Condition 1) of Theorem 5.1 is fulfilled whenever

$$\int_{-\infty}^{\infty} \|d[\tau(\lambda) - \tau_0(\lambda)]\| < \infty. \tag{5.17}$$

10.6. On the limit Weyl-Titchmarsh function

10.6.1. In this section we continue our investigation of the system (1.17). In particular, we will derive a formula for the *limit Weyl-Titchmarsh function* of this system. It follows from Proposition 3.3 that the system (1.17) is in the Weyl limit-point case. Setting

$$R_1(z) = I_m, \qquad Q_1(z) = 0 \tag{6.1}$$

in (2.27), we obtain the following formula for the limit Weyl-Titchmarsh matrix function:

$$v_+(z) = \lim_{l \to +\infty} i\, a_1(l,z)\, c_1^{-1}(l,z), \tag{6.2}$$

where $a_1(l,z)$ and $c_1(l,z)$ are defined by relation (2.28).

10.6.2. Together with $P_1(x,z)$, $P_2(x,z)$ (see (1.14), (1.15)) let us introduce the matrix-valued functions

$$\widehat{P}_1(x,z) = W_{11}(x,z)S_{11}(0) + W_{12}(x,z)S_{21}(0), \tag{6.3}$$

$$\widehat{P}_2(x,z) = W_{21}(x,z)S_{11}(0) + W_{22}(x,z)S_{21}(0). \tag{6.4}$$

Then relation (2.28) can be written in the form

$$\mathfrak{A}_1(l,z) = \begin{bmatrix} \widehat{P}_1^*(l,\bar{z}) & \widehat{P}_2^*(l,\bar{z}) \\ P_1^*(l,\bar{z}) & P_2^*(l,\bar{z}) \end{bmatrix}. \tag{6.5}$$

Consequently, relation (6.2) becomes

$$v_+(z) = \lim_{l \to +\infty} i\, \widehat{P}_1^*(l,\bar{z})\, P_1^*(l,\bar{z})^{-1}, \qquad \operatorname{Im} z > 0. \tag{6.6}$$

Remark 6.1. The uniqueness of the limit Weyl-Titchmarsh function implies the uniqueness of the spectral function $\tau(u)$ corresponding to the boundary value problem (1.17), (1.21) on the half-line. For the case $m = 1$, $d_1 = 1$ this fact is well known [48].

Proposition 6.1. *The following relation holds:*

$$v_+(z) = -i \lim_{l \to +\infty} P_2^{-1}(l,z)\widehat{P}_2(l,z), \qquad \operatorname{Im} z > 0. \tag{6.7}$$

Proof. From (2.28) it follows that

$$\mathfrak{A}_1^*(l,\bar{z})J\mathfrak{A}_1(l,z) = j, \tag{6.8}$$

which in turn yields

$$\widehat{P}_1^*(l,\bar{z})P_1^*(l,\bar{z})^{-1} = -P_2^{-1}(l,z)\, \widehat{P}_2(l,z). \tag{6.9}$$

The needed relation (6.7) is a direct consequence of (6.6) and (6.9). □

10.6.3. By (1.4), the matrix-valued functions $\widehat{P}_1(x,z)$ and $\widehat{P}_2(x,z)$, too, are solutions of the system (1.17). Hence, if we replace $\widehat{P}_k(x,z)$ for $P_k(x,z)$ $(k=1,2)$ all the results of Section 1 remain valid. Let us introduce the following limits:

$$\mathcal{P}(z) = \lim_{x\to\infty} P_2(x,z), \qquad \widehat{\mathcal{P}}(z) = \lim_{x\to\infty} \widehat{P}_2(x,z). \tag{6.10}$$

Proposition 6.2. *If one of limits* (6.10) *exists, then so does the other, and the following equality holds:*

$$v_+(z) = -i\,\mathcal{P}^{-1}(z)\widehat{\mathcal{P}}(z), \qquad \operatorname{Im} z > 0. \tag{6.11}$$

This assertion is a direct consequence of equalities (6.7) and (6.10).

10.6.4. Let us consider separately the case when

$$\int_0^\infty \|A(x)\|\,dx < \infty. \tag{6.12}$$

Theorem 6.1. *Let condition* (6.12) *be satisfied. Then the following assertions hold true:*

1) *The function $v_+(z)$ is continuous in the closed upper half-plane* $\operatorname{Im} z \geq 0$ *and*

$$\|v_+(z)\| \leq M, \qquad \|v_+^{-1}(z)\| \leq M, \qquad \operatorname{Im} z \geq 0, \tag{6.13}$$

where

$$M \leq \|S_{22}(0)\|^2 \exp\left(2\int_0^\infty \|A_{21}(x)\|\,dx\right). \tag{6.14}$$

2) *The spectral matrix function $\tau(u)$ of system* (1.17), (1.21) *is continuously differentiable and*

$$\tau'(u) = \frac{1}{2\pi}\,\mathcal{P}^{-1}(u)\mathcal{P}^{*-1}(u). \tag{6.15}$$

Proof. By Corollary 1.1, the matrix-valued functions $\mathcal{P}(z)$ and $\widehat{\mathcal{P}}(z)$ are continuous in the domain $\operatorname{Im} z \geq 0$. Since $\det \mathcal{P}(z) \neq 0$, the matrix-valued function $v_+(z)$ is also continuous in the domain $\operatorname{Im} z \geq 0$. The estimates (6.13) and (6.14) follow from Lemma 1.2. In order to prove (6.15), we use equality (6.8) for $z = u = \bar{u}$:

$$\mathfrak{A}_1^*(l,u)J\mathfrak{A}_1(l,u) = j, \tag{6.16}$$

which in turn yields

$$P_2(l,u)\widehat{P}_2^*(l,u) + \widehat{P}_2(l,u)P_2^*(l,u) = -I_m. \tag{6.17}$$

Letting here $l \to \infty$, we obtain

$$\mathcal{P}(u)\widehat{\mathcal{P}}^*(u) + \widehat{\mathcal{P}}(u)\mathcal{P}^*(u) = -I_m. \tag{6.18}$$

Using relations (6.11) and (6.18), we obtain the equality

$$\frac{v_+(u) - v_+^*(u)}{i} = \mathcal{P}^{-1}(u)\mathcal{P}^{*-1}(u). \tag{6.19}$$

Finally, relation (6.15) follows from equalities (2.15), (6.19) and the inverse Stieltjes formula. The theorem is proved. □

Chapter 11
Classical Spectral Problems

In this chapter we apply the general spectral theory developed in Chapter 4 to the investigation of a particular class of canonical systems, which includes the string equation and the Sturm-Liouville equation as special cases.

11.1. Generalized string equation (direct spectral problem)

11.1.1. Let us consider the canonical system of equations

$$\frac{d\mathcal{W}}{dx} = izJ\mathcal{H}(x)\mathcal{W}, \qquad \mathcal{W}(0,z) = I_{2m}. \tag{1.1}$$

We shall assume that the Hamiltonian $\mathcal{H}(x)$ is continuous and belongs to the class $\mathfrak{N}(M_2)$, i.e., the matrix $J\mathcal{H}(x)$ is linearly similar to the matrix

$$M_2 = \begin{bmatrix} 0 & 0 \\ I_m & 0 \end{bmatrix}. \tag{1.2}$$

This means that the matrix $\mathcal{H}(x)$ has the following block structure

$$J\mathcal{H}(x) = \begin{bmatrix} p(x) \\ q(x) \end{bmatrix} \left[q^*(x) \;\; p^*(x) \right], \tag{1.3}$$

where $p(x)$ and $q(x)$ are continuous $m \times m$ matrix-valued functions and the following relations hold:

$$q^*(x)p(x) + p^*(x)q(x) = 0, \qquad \operatorname{rank}\mathcal{H}(x) = m. \tag{1.4}$$

In what follows we assume that the normalization condition

$$p(x) > 0 \tag{1.5}$$

is satisfied. Under the assumptions (1.3) and (1.4), both the canonical system (1.1) and the corresponding isometric operator V (see Chapter 4) can be written in a simpler form. Indeed, let us introduce the matrix-valued functions

$$\varphi_1(x,z) = p(x)\mathcal{W}_{21}(x,z) + q^*(x)\mathcal{W}_{11}(x,z), \tag{1.6}$$

$$\varphi_2(x,z) = p(x)\mathcal{W}_{22}(x,z) + q^*(x)\mathcal{W}_{12}(x,z). \tag{1.7}$$

It follows from (1.1) that

$$\begin{cases} \dfrac{dW_{1k}}{dx} = izp(x)\varphi_k(x,z), \\ \\ \dfrac{dW_{2k}}{dx} = izq(x)\varphi_k(x,z), \end{cases} \qquad k = 1, 2. \tag{1.8}$$

Integrating this system and using (1.6) and (1.7), we obtain

$$\varphi_1(x,z) = iz \int_0^x [p(x)q(t) + q^*(x)p(t)]\, \varphi_1(t,z)\, dt + q^*(x), \tag{1.9}$$

$$\varphi_2(x,z) = iz \int_0^x [p(x)q(t) + q^*(x)p(t)]\, \varphi_2(t,z)\, dt + p^*(x). \tag{1.10}$$

Note that the functions $\varphi_1(x,z)$ and $\varphi_2(x,z)$ are uniquely defined by equations (1.9) and (1.10), respectively.

Formula (1.18) of Chapter 4 can be recast as

$$F(u) = V_1 f(x) = \int_0^l Y^*(x,u) f(x)\, dx, \tag{1.11}$$

where

$$Y(x,u) = \varphi_1(x,u)D_1^* + \varphi_2(x,u)D_2^* \tag{1.12}$$

and

$$f(x) = q^*(x)g_1(x) + p(x)g_2(x). \tag{1.13}$$

Here it is assumed that the $m \times m$ matrices D_1 and D_2 obey the relations

$$D_1D_2^* + D_2D_1^* = 0, \qquad D_1D_1^* + D_2D_2^* = I_m.$$

According to the general definition (Chapter 4), the spectral matrix function $\tau(u)$ of system (1.1) can be characterized by the relation

$$\int_{-\infty}^{\infty} F^*(u)[d\tau(u)]F(u) = \int_0^l f^*(x)f(x)\, dx$$

and the following identity holds:

$$\int_0^l g^*(x)\mathcal{H}(x)g(x)\, dx = \int_0^l f^*(x)f(x)\, dx.$$

Thus, the operator V_1 maps the space $L_m^2(0,\infty)$ isometrically into $L_m^2(\tau)$. By formulas (1.9), (1.10), and (1.12), $Y(x,z)$ satisfies the equation

$$Y(x,z) = iz \int_0^x [p(x)q(t) + q^*(x)p(t)]\, Y(t,z)\, dt + q^*(x)D_1^* + p(x)D_2^*. \tag{1.14}$$

Equation (1.14) is called the *generalized string equation*.

11.1.2. Under additional requirements the system of integral equations (1.14) can be reduced to a system of second-order differential equations (matrix string equation).

Theorem 1.1. *Let the following conditions be satisfied:*

1) *The matrices* $p(x)$ *and* $q(x)$ *are continuous and obey relations* (1.4) *and* (1.5).
2) *The matrix* $p(x)$ *is invertible.*
3) *The matrix* $p^{-1}(x)q^*(x)$ *is continuously differentiable.*
4) *The matrix*

$$r(x) = \frac{d}{dx}[p^{-1}(x)q^*(x)] \tag{1.15}$$

is invertible.

Then the system of integral equations (1.14) *is equivalent to the system of differential equations*

$$p^{-1}(x)\frac{d}{dx}\left\{A(x)\frac{d}{dx}\left[p^{-1}(x)Y(x,z)\right]\right\} = zY(x,z) \tag{1.16}$$

supplemented by the conditions

$$Y(0,z) = q^*(0)D_1^* + p(0)D_2^*, \tag{1.17}$$

$$\frac{d}{dx}[p^{-1}(x)Y(x,z)]\big|_{x=0} = r(0)D_1^*, \tag{1.18}$$

where

$$A(x) = -ir^{-1}(x). \tag{1.19}$$

Proof. By relations (1.4) and (1.6)–(1.8),

$$\frac{d}{dx}[p^{-1}(x)\varphi_k(x,z)] = r(x)\mathcal{W}_{1k}(x,z). \tag{1.20}$$

Suppose in addition that the matrix-valued function $r(x)$ is invertible. Then it follows from (1.8) and (1.20) that

$$p^{-1}(x)\frac{d}{dx}\left\{A(x)\frac{d}{dx}\left[p^{-1}(x)\varphi_k(x,z)\right]\right\} = z\varphi_k(x,z), \tag{1.21}$$

which in conjunction with (1.12) yields equation (1.16). Conditions (1.17), (1.18) follow from (1.12), (1.20). The theorem is proved. □

Remark 1.1. The differential operator figuring in the right-hand side of (1.16) is formally self-adjoint since, by (1.4), (1.15), and (1.19),

$$A(x) = A^*(x). \tag{1.22}$$

By strengthening the conditions of Theorem 1.1, we can write them in the following simpler form.

Theorem 1.2. *Let the following conditions be satisfied:*

1) *The matrices* $p(x)$ *and* $q(x)$ *are continuously differentiable and satisfy relations* (1.4), (1.5).
2) *The matrices* $p(x)$ *and*

$$r(x) = \frac{d}{dx}[p^{-1}(x)q^*(x)]$$

are invertible.

Then the system of integral equations (1.14) *is equivalent to the differential system* (1.16) *supplemented by conditions* (1.17), (1.18).

11.1.3. As the reader is probably aware of, the system (1.16) is itself of great interest. Therefore, it is important to formulate the above results for the system (1.16), i.e., in terms of the matrices $A(x)$ and $p(x)$.

Theorem 1.3. *Let the following conditions be fulfilled.*

1) *The matrices* $p(x)$ *and* $A(x)$ *are continuous and invertible.*
2) *Relations* (1.5) *and* (1.22) *hold.*

Then the differential system (1.16)–(1.18) *is equivalent to the system of integral equations* (1.14), *where*

$$q^*(x) = p(x)\left[\int_0^x r(t)\,dt + r(0)\right], \tag{1.23}$$

$$r(x) = -iA^{-1}(x). \tag{1.24}$$

Proof. It is readily seen that $p(x)$ and $q(x)$ satisfy all the conditions of Theorem 1.1. □

Setting

$$Y_1(x,z) = p^{-1}(x)Y(x,z) \tag{1.25}$$

we can rewrite (1.16) in the form

$$\frac{d}{dx}\left\{A(x)\left[\frac{d}{dx}Y_1(x,z)\right]\right\} = zp^2(x)Y_1(x,z). \tag{1.26}$$

Equation (1.26) is a matrix generalization of the classical string equation.

Example 1.1. Suppose that

$$p(x) = I_m, \qquad q(x) = -ixI_m, \qquad D_1 = 0, \qquad D_2 = I_m. \tag{1.27}$$

Then

$$A(x) = -I_m, \qquad Y(x,z) = \cos(\sqrt{z}x)I_m. \tag{1.28}$$

Taking

$$D_1 = 0, \qquad D_2 = I_m \tag{1.29}$$

let us apply the general spectral theorems of Chapter 4 to system (1.16).

Remark 1.2. The transition from case (1.29) to the general case is explained in Chapter 4.

The following assertion is a consequence of Theorem 3.3 of Chapter 4.

Theorem 1.4. *Let the $m \times m$ matrices $p(x)$ and $q(x)$ be continuous and satisfy relations (1.4) and (1.5). Suppose $\lambda = 0$ is not an eigenvalue of the operator A_M, acting in $L_2^m(0,l)$ as*

$$A_M f = i\int_0^x [p(x)q(t) + q^*(x)p(t)]f(t)\,dt. \tag{1.30}$$

Then the set of the Weyl-Titchmarsh matrix functions $v(z)$ of the corresponding system (1.14), (1.29) is described by the formula

$$v(z) = i[a(z)\mathcal{P}(z) + b(z)\mathcal{Q}(z)][c(z)\mathcal{P}(z) + d(z)\mathcal{Q}(z)]^{-1}, \tag{1.31}$$

where $\mathcal{P}(z)$, $\mathcal{Q}(z)$ is an arbitrary non-special pair with the J-property and the matrix of coefficients has the form

$$\mathfrak{A}(l,z) = \begin{bmatrix} a(z) & b(z) \\ c(z) & d(z) \end{bmatrix} = \mathcal{W}^*(l,\bar{z}). \tag{1.32}$$

(*The matrix $\mathcal{W}(l,z)$ is defined by* (1.1).)

Recall that the spectral data $\tau(u)$ and $\alpha = \alpha^*$ of system (1.14) are connected with its Weyl-Titchmarsh matrix function by the equality

$$v(z) = \alpha + \int_{-\infty}^{\infty}\left(\frac{1}{u-z} - \frac{u}{1+u^2}\right)d\tau(u). \tag{1.33}$$

Thus, Theorem 1.4 gives a solution of the direct spectral problem for the system (1.14), (1.29).

Remark 1.3. In the case (1.29), conditions (1.17) and (1.18) have the form

$$Y(0,z) = p(0), \qquad \frac{d}{dx}[p^{-1}(x)\, Y(x,z)]|_{x=0} = 0. \tag{1.34}$$

11.1.4. If the conditions of Theorem 1.2 are satisfied, then so are the conditions of Theorem 1.4, i.e., the following assertion holds true.

Theorem 1.5. *Let the conditions of Theorem* 1.2 *be satisfied. Then the set of the Weyl-Titchmarsh matrix functions* $v(z)$ *of the corresponding system* (1.16), (1.34) *is described by formulas* (1.31), (1.32).

Proof. Assume that $\lambda = 0$ is an eigenvalue of the operator A_{M}. Then the equality

$$\int_0^x [p(x)q(t) + q^*(x)p(t)]\, f_0(t)\, dt = 0 \tag{1.35}$$

holds for some $f_0 \in L^2_m(0,l)$ such that $\|f_0\| \neq 0$.

Denote by $H_{1,x}$ the space spanned by the rows of the matrix $[p(x) \;\; q^*(x)]$ and by $H_{2,x}$ the space spanned by the rows of the matrix $[q^*(x) \;\; p(x)]$. Thanks to (1.4), these spaces are orthogonal to one another and are both of dimension m. It follows from (1.35) that the $1 \times 2m$ vector

$$g_0(x) = \left[\int_0^x f_0^*(t) q^*(t)\, dt \quad \int_0^x f_0^*(t) p(t)\, dt\right]$$

belongs to $H_{2,x}$. Hence, there exists an $1 \times m$ vector $h_0(x)$ such that

$$h_0(x)[q^*(x) \;\; p(x)] = g_0(x),$$

i.e.,

$$h_0(x)q^*(x) = \int_0^x f_0^*(t)q^*(t)\, dt \tag{1.36}$$

and

$$h_0(x)p(x) = \int_0^x f_0^*(t)p(t)\, dt. \tag{1.37}$$

Since the matrices $p(x)$ and $q^*(x)$ are invertible and differentiable, relations (1.36) and (1.37) show that $h_0(x)$ is absolutely continuous. Hence, for almost every x,

$$h_0'(x)q^*(x) + h_0(x)q^{*\prime}(x) = f_0^*(x)q^*(x), \tag{1.38}$$

$$h_0'(x)p(x) + h_0(x)p'(x) = f_0^*(x)p(x). \tag{1.39}$$

Multiplying to the left both sides of (1.38) by $p(x)$ and both sides of (1.39) by $q(x)$ and adding the resulting expressions yields

$$h_0(x)r_1(x) = 0, \tag{1.40}$$

where

$$r_1(x) = q^{*\prime}(x)p(x) + p'(x)q(x). \tag{1.41}$$

Comparing (1.15) and (1.41), we conclude that

$$r(x) = p^{-1}(x)r_1(x)p^{-1}(x). \tag{1.42}$$

Further, from the invertibility of the matrix $r(x)$ (condition 2) of Theorem 1.2) it follows that the matrix $r_1(x)$ is also invertible. Then, by (1.40), $h_0(x) = 0$, i.e. $f_0(x) = 0$. The contradiction we have reached shows that $\lambda = 0$ is not an eigenvalue of the operator A_M. We see that all the conditions of Theorem 1.4 are satisfied, which completes the proof of Theorem 1.5. □

Corollary 1.1. *Let the conditions of Theorem* 1.1 *be satisfied and let the matrix-valued function $p(x)$ be continuously differentiable. Then the assertion of Theorem* 1.5 *holds true.*

A survey of results on the spectral problem for the scalar string equation is given in the paper by I. S. Kats and M. G. Kreĭn [40] (see also [18]–[20]).

11.2. Matrix Sturm-Liouville equation (direct spectral problem)

11.2.1. Let $U(x)$ be a $m \times m$ matrix-valued function and let

$$U(x) = U^*(x). \tag{2.1}$$

We introduce the matrix-valued function

$$G(x,z) = \begin{bmatrix} 0 & I_m \\ U(x) - zI_m & 0 \end{bmatrix} \tag{2.2}$$

and consider the following system of differential equations:

$$\frac{dW}{dx} = G(x,z)W, \qquad 0 \le x \le l \le \infty, \qquad W(0,z) = I_{2m}. \tag{2.3}$$

Further, let

$$P = \begin{bmatrix} 0 & I_m \\ U & 0 \end{bmatrix}, \qquad j_1 = i\begin{bmatrix} 0 & -I_m \\ I_m & 0 \end{bmatrix}, \qquad T_1 = \frac{1}{\sqrt{2}}\begin{bmatrix} iI_m & I_m \\ iI_m & -I_m \end{bmatrix}, \tag{2.4}$$

and define the matrix-valued function $B(x)$ as the solution of the system

$$\frac{dB}{dx} = PB, \qquad B(0) = T_1. \tag{2.5}$$

We will assume here that the entries of $U(x)$ are locally integrable. Then the matrix-valued function

$$\mathcal{W}(x,z) = B^{-1}(x)W(x,z)T_1 \tag{2.6}$$

is a solution of the system

$$\frac{d\mathcal{W}}{dx} = izJ\mathcal{H}(x)\mathcal{W}, \qquad \mathcal{W}(0,z) = I_{2m}, \tag{2.7}$$

where

$$J\mathcal{H}(x) = B^{-1}(x)\begin{bmatrix} 0 & 0 \\ iI_m & 0 \end{bmatrix} B(x). \tag{2.8}$$

From (2.4), (2.5) and the equality

$$j_1 P j_1 = -P^* \tag{2.9}$$

we deduce that

$$B^*(x)j_1B(x) = J, \qquad T_1^* j_1 T_1 = J. \tag{2.10}$$

Further, by (2.8) and (2.10),

$$\mathcal{H}(x) = B^*(x)\begin{bmatrix} I_m & 0 \\ 0 & 0 \end{bmatrix} B(x). \tag{2.11}$$

The operator V defined by rule (1.18) of Chapter 4 has the form

$$F(u) = Vg(x) = \int_0^l [D_1 \ \ D_2]T^*W^*(x,u)\mathrm{col}[f_1(x), 0]\, dx, \tag{2.12}$$

where

$$f(x) = \mathrm{col}[f_1(x), f_2(x)] = A(x)g(x). \tag{2.13}$$

Let us introduce $Y_1(x,z)$, $Y_2(x,z)$, and $Y(x,z)$ by the relations

$$Y_1(x,z) = \frac{i}{\sqrt{2}}(W_{11}(x,z) + W_{12}(x,z)), \tag{2.14}$$

$$Y_2(x,z) = \frac{1}{\sqrt{2}}(W_{11}(x,z) - W_{12}(x,z)), \tag{2.15}$$

and

$$Y(x,z) = Y_1(x,z)D_1^* + Y_2(x,z)D_2^*, \tag{2.16}$$

respectively, where $W_{ij}(x,z)$ are the $m \times m$ blocks of the matrix $W(x,z)$. Then (2.12) can be written in the form

$$F(u) = V_1 f_1(x) = \int_0^l Y^*(x,u) f_1(x)\, dx. \tag{2.17}$$

Thus, the operator V_1 maps $L^2_m(0,\infty)$ isometrically into $L^2_m(\tau)$. Formulas (2.2), (2.3) and (2.14)–(2.16) imply that $Y(x,z)$ satisfies the matrix Sturm-Liouville equation

$$-\frac{d^2Y}{dx^2}+U(x)Y(x,z)=zY(x,z) \tag{2.18}$$

and the conditions

$$Y(0,z)=\frac{1}{\sqrt{2}}(iD_1^*+D_2^*),\qquad Y'(0)=\frac{1}{\sqrt{2}}(iD_1^*-D_2^*). \tag{2.19}$$

The boundary conditions (2.19) can be recast as

$$C_1Y(0)=C_2Y'(0), \tag{2.20}$$

where

$$C_1=D_1-iD_2,\qquad C_2=D_1+iD_2. \tag{2.21}$$

Here we have used the relations

$$D_1D_2^*+D_2D_1^*=0,\qquad D_1D_1^*+D_2D_2^*=I_m. \tag{2.22}$$

The following assertion holds true.

Theorem 2.1. *The spectral matrix function $\tau(u)$ introduced above coincides with the classical spectral matrix function of the Sturm-Liouville problem* (2.18), (2.20).

Remark 2.1. Thanks to (2.22), the following equality holds:

$$C_1C_2^*=C_2C_1^*. \tag{2.23}$$

Consequently, the operator defined by the differential expression (2.18) and the boundary conditions (2.20) admits a self-adjoint extension.

11.2.2. As in Subsection 11.1, let us set

$$D_1=0,\qquad D_2=I_m. \tag{2.24}$$

Then condition (2.20) takes the form

$$Y(0)=-Y'(0). \tag{2.25}$$

Let us apply the general spectral theorems (Chapter 4) to the system (2.18), (2.25). To this end we consider the block decomposition of the matrix $B(x)$:

$$B(x)=\begin{bmatrix}B_{11}(x) & B_{12}(x)\\ B_{21}(x) & B_{22}(x)\end{bmatrix},$$

where all the blocks are of order $m\times m$.

Theorem 2.2. *Let the $m \times m$ matrix-valued function $U(x)$ satisfy the conditions*

$$U(x) = U^*(x), \qquad \int_0^l \|U(x)\|\, dx < \infty. \tag{2.26}$$

Then the set of the Weyl-Titchmarsh matrix functions $v(z)$ of the system

$$-\frac{d^2Y}{dx^2} + U(x)Y = zY, \qquad Y(0) = -Y'(0) \tag{2.27}$$

is described by the formula

$$v(z) = i[a(z)\mathcal{P}(z) + b(z)\mathcal{Q}(z)][c(z)\mathcal{P}(z) + d(z)\mathcal{Q}(z)]^{-1}, \tag{2.28}$$

where $\mathcal{P}(z)$, $\mathcal{Q}(z)$ is an arbitrary non-special pair with the J-property and the matrix of coefficients has the form

$$\mathfrak{A}(l,z) = \begin{bmatrix} a(z) & b(z) \\ c(z) & d(z) \end{bmatrix} = \mathcal{W}^*(l, \bar{z}). \tag{2.29}$$

(The matrix $\mathcal{W}(l,z)$ is defined by (2.7).*)*

Proof. Rewrite formula (2.11) in the form

$$\mathcal{H}(x) = \beta^*(x)\beta(x), \tag{2.30}$$

where

$$\beta(x) = [\beta_1(x) \;\; \beta_2(x)], \tag{2.31}$$

$$\beta_1(x) = q^*(x) = B_{11}(x), \qquad \beta_2(x) = p^*(x) = B_{22}(x). \tag{2.32}$$

Now consider the *model operator* A_{M}, acting according to the rule

$$A_{\mathrm{M}} f = i\beta(x)J \int_0^x \beta^*(t) f(t)\, dt = i\int_0^x [p^*(x)q(t) + q^*(x)p(t)] f(t)\, dt. \tag{2.33}$$

It follows from relation (2.5) that

$$\frac{dB_{11}}{dx} = B_{21}, \qquad \frac{dB_{12}}{dx} = B_{22}, \qquad \frac{dB_{21}}{dx} = UB_{11}, \qquad \frac{dB_{22}}{dx} = UB_{12}. \tag{2.34}$$

This implies that the matrix-valued functions $q(x)$ and $p(x)$ are twice differentiable and

$$\frac{d^2p^*(x)}{dx^2} = U(x)p^*(x), \tag{2.35}$$

$$\frac{d^2q^*(x)}{dx^2} = U(x)q^*(x). \tag{2.36}$$

Next, by (2.10),

$$B(x)JB^*(x) = j_1. \tag{2.37}$$

Consequently

$$B_{12}B_{11}^* + B_{11}B_{12}^* = 0 \tag{2.38}$$

and

$$B_{21}B_{12}^* + B_{22}B_{11}^* = iI_m. \tag{2.39}$$

Using (2.32) and (2.34), formulas (2.38) and (2.39) can be written as

$$p^*(x)q(x) + q^*(x)p(x) = 0 \tag{2.40}$$

and

$$q'^*(x)p(x) + p'^*(x)q(x) = iI_m, \tag{2.41}$$

respectively. Differentiating (2.33) twice and taking into account (2.40), (2.41), we obtain

$$(A_{\mathrm{M}}f)'' = -f(x) + i\int_0^x \left[p''^*(x)q(t) + q''^*(x)p(t)\right] f(t)\,dt \neq 0 \qquad (\|f\| \neq 0),$$

i.e., $\lambda = 0$ is not an eigenvalue of A_{M}. Finally, by (2.4) and (2.5),

$$p(0) = \frac{1}{\sqrt{2}} I_m.$$

Thus all the conditions of Theorem 3.3 of Chapter 4 are satisfied. The theorem is proved. □

The spectral data $\tau(u)$, α of system (2.18), (2.25) are connected with the corresponding Weyl-Titchmarsh matrix functions $v(z)$ by equality (1.33). Thus, Theorem 2.2 gives a solution of the direct spectral problem for the matrix Sturm-Liouville problem (2.27).

11.3. Inverse spectral problem

11.3.1. Let us apply the general procedure of solving the inverse spectral problem (Chapter 4) to the following example.

Consider the operator A acting as

$$Af = \int_0^x (t-x)f(t)\,dt, \qquad 0 \leq x \leq l, \tag{3.1}$$

where $f(x) \in L_m^2(0, l)$. Recall that the projectors P_ζ are defined by the rule

$$P_\zeta f = \begin{cases} f, & \text{for } 0 \leq x \leq \zeta \\ 0 & \text{for } x > \zeta. \end{cases} \tag{3.2}$$

Let us begin with the simplest case in which

$$S_0 = I. \tag{3.3}$$

In this case we have

$$(A - A^*)f = \int_0^l (t - x)f(t)\,dt, \tag{3.4}$$

which can be recast as the operator identity

$$A - A^* = i(\Phi_1\Phi_2^* + \Phi_2\Phi_1^*),$$

were

$$\Phi_2 g = g, \qquad \Phi_1 g = ixg. \tag{3.5}$$

Here g are constant $m \times 1$ vectors.

Using the formula

$$\mathcal{H}_0(\zeta) = \frac{d}{d\zeta}\left(\Pi^* S_\zeta^{-1} P_\zeta \Pi\right), \qquad \Pi = [\Phi_1 \;\; \Phi_2]$$

(see Chapter 4, Section 3) we obtain the relation

$$J\mathcal{H}_0(x) = \begin{bmatrix} ixI_m & I_m \\ x^2 I_m & -ixI_m \end{bmatrix}. \tag{3.6}$$

It follows from (3.4) and (3.5) that

$$(I - A\lambda)^{-1}\Phi_2 = \cos(\sqrt{\lambda}x)\Phi_2. \tag{3.7}$$

Hence, the operator $S_0 = I$ admits the representation

$$S_0 = \int_{-\infty}^{\infty} (I - A\lambda)^{-1}\Phi_2[d\tau_0(\lambda)]\Phi_2^*(I - A^*\lambda)^{-1}, \tag{3.8}$$

where

$$\tau_0(\lambda) = \begin{cases} 0, & \text{for } \lambda \le 0 \\ \frac{2}{\sqrt{\pi}}\sqrt{\lambda} & \text{for } \lambda > 0. \end{cases}$$

From (3.8) we deduce that $\tau(\lambda)$ is a spectral matrix function of the system

$$\frac{dY}{dx} = izJ\mathcal{H}_0(x)Y, \qquad 0 \le x < \infty. \tag{3.9}$$

Applying the general spectral theory (Chapter 4) we obtain the following result.

Theorem 3.1. *Let the operators A, Φ_2 and P_ζ be defined by formulas* (3.1), (3.2) *and* (3.5), *respectively. Suppose that the monotonically increasing matrix-valued function $\tau(\lambda)$ satisfies the following conditions:*

1) *The matrix-valued function*

$$K(x,u) = \int_{-\infty}^{\infty} \cos\sqrt{\lambda}x\,\{d[\tau(\lambda) - \tau_0(\lambda)]\}\cos(\sqrt{\lambda}u), \qquad 0 \le x,\, u < \infty \tag{3.10}$$

is continuous.

2) *If*

$$\int_{-\infty}^{\infty} F^*(\lambda)[d\tau(\lambda)]F(\lambda) = 0$$

then $F(\lambda) = 0$, where one denotes

$$F(\lambda) = \int_0^l \cos(\sqrt{\lambda}x) f(x)\,dx. \tag{3.11}$$

Then $\tau(\lambda)$ is a spectral matrix function of the system (1.1). *The corresponding Hamiltonian $\mathcal{H}(x)$ is continuous and can be found via the following chain of formulas:*

$$S_\zeta f = f(x) + \int_0^\zeta K(x,u) f(u)\,du, \tag{3.12}$$

$$\mathcal{H}(\zeta) = \frac{d}{d\zeta}(\Pi^* S_\zeta^{-1} P_\zeta \Pi), \qquad \Pi = [\Phi_1 \ \ \Phi_2], \tag{3.13}$$

$$\Phi_1 = -i\int_{-\infty}^{\infty} \frac{\cos(\sqrt{\lambda}x) - 1}{\lambda(1+\lambda^2)}\,\Phi_2\,d\tau(\lambda). \tag{3.14}$$

Proof. The general formula for the operator S (Chapter 4) reads

$$S = \int_{-\infty}^{\infty} (I - A\lambda)^{-1}\Phi_2[d\tau(\lambda)]\Phi_2^*(I - A^*\lambda)^{-1}. \tag{3.15}$$

Since in the case under consideration equality (3.7) holds, relations (3.2) and (3.10) yield equality (3.12) with $S_\zeta = P_\zeta S P_\zeta$.

Further, the general formula for Φ_1 (Chapter 4) reads

$$\Phi_1 = -i\int_{-\infty}^{\infty} \left[A(I - A\lambda)^{-1} + \frac{\lambda}{1+\lambda^2} I\right]\Phi_2\,d\tau(\lambda). \tag{3.16}$$

In the case (3.1), (3.2) formula (3.16) coincides with (3.14). It follows from representation (3.15) that $S \ge 0$. Thanks to condition 2) of the theorem,

$$S \ge \delta I \tag{3.17}$$

for some $\delta > 0$. From (3.12) and (3.17) we obtain the existence and boundedness of the operator S_ζ^{-1}. Hence, formula (3.13) is valid. By formulas (2.13) and (2.14) of Chapter 7, the Hamiltonian $\mathcal{H}(x)$ is continuous. This completes the proof of the theorem. □

Remark 3.1. Condition 1) of Theorem 3.1 is satisfied whenever

$$\int_0^\infty \|d\,[\tau(\lambda) - \tau_0(\lambda)]\,\| + \int_0^\infty e^{\sqrt{|\lambda|}x}\|d\tau(\lambda)\| < \infty. \tag{3.18}$$

It follows from (3.6) that the matrix $J\mathcal{H}_0(x)$ is linearly similar to the matrix

$$M_2 = \begin{bmatrix} 0 & 0 \\ I_m & 0 \end{bmatrix}.$$

Hence, Corollary 2.2 of Chapter 7 admits the following consequence

Corollary 3.1. *Let the conditions of Theorem* 3.1 *be satisfied. Then the corresponding matrix* $\mathcal{H}(x)$ *has the property that*

$$\operatorname{rank} \mathcal{H}(x) \le m.$$

If $\operatorname{rank} \mathcal{H}(x) = m$ *then the matrix* $J\mathcal{H}(x)$ *is linearly similar to the matrix* M_2.

From this corollary it further follows that the matrix $J\mathcal{H}(x)$ admits the representation (1.3), (1.4).

Thus, Theorem 3.1 allows us to reconstruct the canonical system (1.1) with a Hamiltonian of the form (1.3), (1.4) from its spectral matrix function $\tau(\lambda)$.

Chapter 12
Nonlinear Integrable Equations and the Method of the Inverse Spectral Problem

Nonlinear partial differential and differential-difference equations play an important role in many problems of physics. For a long period of time the main method for studying such equations was to consider them, whenever possible, as small perturbations of their main, linear part. However, in 1967, C. S. Gardner, J. M. Green, M. D. Kruskal, and R. M. Miura [23] published a paper in which they introduced a fundamentally different approach to the analysis of the nonlinear Korteweg-de Vries equation

$$U_t - 6UU_x + U_{xxx} = 0. \tag{0.1}$$

Specifically, they used the results of the inverse scattering problem for the equation

$$-\frac{d^2y}{dx^2} + U(x,t)y = ly, \qquad -\infty < x < \infty.$$

The method of the inverse scattering problem was further developed by P. D. Lax [54] and by V. E. Zakharov and A. B. Shabat [98]. It was shown that this method can be applied not only to the Korteweg-de Vries equation, but also to a large number of other nonlinear equations of mathematical physics. The method of the inverse scattering problem has been widely applied by numerous investigators (see, e.g. [1], [15]) and continues to be a very active field of research.

The method of the inverse spectral problem was for the first time applied to the analysis of nonlinear equations by Yu. M. Berezanskiĭ [11] (in the discrete case) and by the author of this book [86], [89] (in discrete and continuous cases).

The transition from the inverse scattering problem to the inverse spectral problem removes the requirement of regularity of solutions at infinity and allows one to construct new classes of explicit solutions for a number of nonlinear partial differential equations.

Chapter 12 is devoted to the description of the main ideas of the method of the inverse spectral problem.

12.1. Evolution of spectral data

12.1.1. Let $G(x,t,z)$, $F(x,t,z)$ be $n \times n$ matrix-valued functions continuously differentiable in the domain $0 \le x, t < \infty$. Denote by $\mathcal{W}(x,t,z)$ the solution of the equation

$$\frac{\partial \mathcal{W}}{\partial x} = G(x,t,z)\mathcal{W}, \qquad 0 \le x,\, t < \infty \tag{1.1}$$

satisfying the condition

$$\mathcal{W}(0,t,z) = I_n. \tag{1.2}$$

The following assertion can be verified directly.

Proposition 1.1. *The matrix-valued function*

$$U(x,t,z) = \frac{\partial \mathcal{W}}{\partial t} - F(x,t,z)\mathcal{W} \tag{1.3}$$

satisfies equation (1.1) *if and only*

$$\frac{\partial G}{\partial t} - \frac{\partial F}{\partial x} + [G,F] = 0. \tag{1.4}$$

(*Here* $[G,F] = GF - FG$.)

Proof. Differentiating relation (1.3) with respect to x and taking into account (1.1), we have

$$\frac{\partial U(x,t,z)}{\partial x} = \left(\frac{\partial G}{\partial t} - \frac{\partial F}{\partial x} + GF - FG\right)\mathcal{W} + GU,$$

i.e., (1.4) holds. □

As is known [98], [22], a number of nonlinear partial differential equations can be written in the form (1.4). Proposition 1.1 connects the nonlinear equation (1.4) with linear system (1.1). From Proposition 1.1 and the relation

$$U(0,t,z) = -F(0,t,z)$$

we deduce the equality

$$\frac{\partial \mathcal{W}}{\partial t} - F(x,t,z)\mathcal{W} = -\mathcal{W}F(0,t,z). \tag{1.5}$$

Therefore, the following relation holds:

$$\mathcal{W}(x,t,z) = V(x,t,z)\mathcal{W}(x,0,z)V^{-1}(0,t,z), \tag{1.6}$$

where $V(x,t,z)$ is a solution of the equation

$$\frac{\partial V}{\partial t} = F(x,t,z)V, \qquad V(x,0,z) = I_n. \tag{1.7}$$

Let us consider separately the important special case when

$$n = 2m, \qquad G(x,t,z) = izJ\mathcal{H}(x,t), \qquad \mathcal{H}(x,t) \geq 0. \tag{1.8}$$

To this end, let us decompose the following matrices into $m \times m$ blocks:

$$V^*(l,t,\bar{z}) = [q_{ij}(l,t,z)]_{i,j=1}^2, \qquad V^{*-1}(0,t,\bar{z}) = [r_{ij}(t,z)]_{i,j=1}^2, \tag{1.9}$$

$$\mathcal{W}^*(l,t,z) = \begin{bmatrix} a(l,t,z) & b(l,t,z) \\ c(l,t,z) & d(l,t,z) \end{bmatrix}, \tag{1.10}$$

$$\mathcal{W}^{*-1}(l,t,\bar{z})J\mathcal{W}^{-1}(l,t,\bar{z}) = \begin{bmatrix} -\gamma(l,t,z) & \delta(l,t,z) \\ \delta^*(l,t,z) & -\nu(l,t,z) \end{bmatrix}. \tag{1.11}$$

Relations (1.1), (1.2), and (1.8) yield the identity

$$\int_0^l \mathcal{W}^*(x,z)[J\mathcal{H}(x)]\mathcal{W}(x,z)\,dx = \frac{i}{z-\bar{z}}[J - \mathcal{W}^*(l,z)J\mathcal{W}(l,z)]. \tag{1.12}$$

(Here the parameter t is omitted.) A consequence of (1.12) is that

$$J - \mathcal{W}^*(l,t,z)J\mathcal{W}(l,t,z) \geq 0 \qquad \text{for} \quad \operatorname{Im} z > 0. \tag{1.13}$$

Hence, $\gamma(l,t,z) \geq 0$ for $\operatorname{Im} z > 0$. If strict inequality holds, i.e.,

$$\gamma(l,t,z) > 0, \qquad \operatorname{Im} z > 0, \tag{1.14}$$

then the following linear-fractional transformation is well defined (see [94]):

$$v(l,t,z) = i\,[a(l,t,z)\mathcal{P}(z) + b(l,t,z)\mathcal{Q}(z)]\cdot[c(l,t,z)\mathcal{P}(z) + d(l,t,z)\mathcal{Q}(z)]^{-1} \tag{1.15}$$

Here $\mathcal{P}(z)$, $\mathcal{Q}(z)$ is an arbitrary non-singular pair of meromorphic matrix-valued functions with the J-property. The matrices $v(l,t,z)$ form a system of *nested Weyl matrix discs*. If, as $l \to \infty$, the Weyl discs contract to a point, then there exists the limit

$$v(t,z) = \lim_{l\to\infty} v(l,t,z), \qquad \operatorname{Im} z > 0. \tag{1.16}$$

The following theorem describes the behavior of $v(t,z)$ as a function of t ("evolution law").

Theorem 1.1. *Let $G(x,t,z)$ and $F(x,t,z)$, be continuously differentiable $n \times n$ matrix-valued functions ($0 \leq x,t < \infty$) satisfying relations* (1.4) *and* (1.8). *Suppose also that the following requirements are fulfilled:*

a) *There exists an l_0 such that, for all $l > l_0$,*

$$\gamma(l,0,z) > 0, \qquad \operatorname{Im} z > 0. \tag{1.17}$$

b) *There exists the limit*

$$v(z) = v(0,z) = \lim_{l\to\infty} v(l,0,z). \tag{1.18}$$

c) *For every* $t > 0$,

$$\det\{r_{21}(t,z)[-iv(z)] + r_{22}(t,z)\} \not\equiv 0, \qquad \operatorname{Im} z > 0. \tag{1.19}$$

d) *For every* $t > 0$ *and every pair* $\mathcal{P}(z)$, $\mathcal{Q}(z)$ *such that*

$$\mathcal{P}^*(z)\mathcal{Q}(z) + \mathcal{Q}^*(z)\mathcal{P}(z) > 0, \qquad \operatorname{Im} z > 0, \tag{1.20}$$

there exist numbers l_0 *and* z_0 *with* $\operatorname{Im} z_0 > 0$ *such that in some neighborhood of* z_0 *the pair* $\mathcal{P}_1(z)$, $\mathcal{Q}_1(z)$ *defined by the formula*

$$[\mathcal{P}_1 \;\; \mathcal{Q}_1] = V^*(l,t,z)[\mathcal{P} \;\; \mathcal{Q}] \tag{1.21}$$

has the J*-property for* $l \geq l_0$.

Then the limit (1.16) *exists and*

$$v(t,z) = i\{r_{11}(t,z)[-iv(z)] + r_{12}(t,z)\} \cdot \{r_{21}(t,z)[-iv(z)] + r_{22}(t,z)\}^{-1}. \tag{1.22}$$

Proof. Conditions a) and d) guarantee that the linear-fractional transformation

$$v_1(l,t,z) = i\{a(l,0,z)\mathcal{P}_1 + b(l,0,z)\mathcal{Q}_1\} \cdot \{c(l,0,z)\mathcal{P}_1 + d(l,0,z)\mathcal{Q}_1\}^{-1} \tag{1.23}$$

is well defined. Since $v_1(l,t,z) \to v(z)$ when $l \to \infty$, it follows from (1.19) that

$$\det\{r_{21}(t,z)[-iv_1(l,t,z)] + r_{22}(t,z)\} \not\equiv 0, \qquad \operatorname{Im} z > 0$$

provided that l is sufficiently large. Further, using (1.19) and (1.23) we deduce from (1.6) and (1.10) that

$$\begin{aligned} v(l,t,z) =& i\{r_{11}(t,z)[-iv_1(l,t,z)] + r_{12}(t,z)\} \times \\ & \times \{r_{21}(t,z)[-iv_1(l,t,z)] + r_{22}(t,z)\}^{-1}. \end{aligned} \tag{1.24}$$

Now from conditions b), c) and equalities (1.23), (1.24) we obtain the relation

$$\begin{aligned} v(t,z) = \lim_{l\to\infty} v(l,t,z) = i\{r_{11}(t,z)[-iv(z)] + r_{12}(t,z)\} \times \\ \times \{r_{21}(t,z)[-iv(z)] + r_{22}(t,z)\}^{-1}, \end{aligned} \tag{1.25}$$

which holds in a neighbourhood of z_0. The matrices $v(l,t,z)$ belong to the *Nevanlinna class* in the variable z and therefore form a normal family of analytic matrix-valued functions. Hence, (1.25) is valid for all z with $\operatorname{Im} z > 0$. Let us note that the limit in (1.25) does not depend on the choice of the pair $\mathcal{P}$, $\mathcal{Q}$. The theorem is proved. □

Remark 1.1. The matrix-valued function $v(t,\cdot)$ belongs to the Nevanlinna class and hence admits the representation

$$v(t,z) = \beta(t)z + \alpha(t) + \int_{-\infty}^{\infty} \left(\frac{1}{u-z} - \frac{u}{1+u^2} \right) d\tau(u,t), \tag{1.26}$$

where $\beta \geq 0$, $\alpha(t) = \alpha^*(t)$ and $\tau(u,t)$ is monotonically increasing with respect to u.

Under additional assumptions indicated in Chapter 4, one can show that $\beta(t) = 0$, and that $\alpha(t)$ and $\tau(u,t)$ are the spectral characteristics of a canonical system (1.1), (1.8).

Remark 1.2. Assume that the inequality

$$\int_0^l \mathcal{H}(x,t)\, dx > 0, \tag{1.27}$$

holds, i.e., system (1.1), (1.8) is of positive type. Then in (1.13) one has strict inequality (see (2.11) and (2.12) of Chapter 6), which implies (1.14).

Formula (1.22) allows us to solve the following problem.

Problem 1.1. Let the initial data $G(x,0,z) = izJ\mathcal{H}(x,0)$ and the boundary data $F(0,t,z)$ for the nonlinear system (1.4) be given. Find the matrix-valued function $v(t,z)$ that solves the corresponding linear system (1.1).

We will assume that the conditions of Theorem 1.1 are fulfilled. Then the recipe for solving Problem 1.1 is as follows.

1. Given $\mathcal{H}(x,0)$, find the matrix-valued function $v(z)$ corresponding to the system (1.1) with $t = 0$ (see Theorem 2.6 of Chapter 4 and Chapter 9, Section 1).
2. By (1.7) and (1.9), the matrix-valued function $R(t,z)$ satisfies the relations

$$\frac{\partial R(t,z)}{\partial t} = -F^*(0,t,\bar{z})R(t,z), \qquad R(0,z) = I_{2m}. \tag{1.28}$$

   Now solve the linear system (1.28) to find $R(t,z)$, i.e., the coefficients of the linear-fractional transformation (1.22).
3. Finally, use formula (1.22) to calculate $v(t,z)$.

Thus, the evolution of $v(t,z)$ (its dependence on time), which is governed by a nonlinear equation, can be determined by resorting to a linear theory. This fact serves as the basis for the application of the method of the inverse spectral problem to nonlinear equations.

12.2. Some classical nonlinear equations

In this section we will show that for a number of well-known nonlinear partial differential equations all the assumptions of Theorem 1.1 concerning the evolution of spectral are satisfied.

Example 1 (nonlinear Schrödinger equation). Following [98], let us introduce the matrix-valued functions

$$G_1(x,t,z) = j(izI_n - \xi), \tag{2.1}$$

$$F_1(x,t,z) = -i\left[jz^2 + izj\xi - \frac{1}{2}\left(\Omega - \frac{\partial \xi}{\partial x}\right)\right], \tag{2.2}$$

where $n = 2m$ and the matrices ξ and Ω have the following decompositions into $m \times m$ blocks:

$$\xi = \begin{bmatrix} 0 & R(x,t) \\ -R^*(x,t) & 0 \end{bmatrix}, \qquad \Omega = \begin{bmatrix} -RR^* & 0 \\ 0 & R^*R \end{bmatrix}, \tag{2.3}$$

with

$$j = \begin{bmatrix} I_m & 0 \\ 0 & -I_m \end{bmatrix}.$$

It follows from relations (2.1)–(2.3) that

$$\frac{\partial G_1}{\partial t} = -j\frac{\partial \xi}{\partial t}, \qquad \frac{\partial F_1}{\partial x} = zj\frac{\partial \xi}{\partial x} + \frac{i}{2}\frac{\partial}{\partial x}\left(\Omega - \frac{\partial \xi}{\partial x}\right), \tag{2.4}$$

and

$$[G_1, F_1] = zj\frac{\partial \xi}{\partial x} + \frac{i}{2}\frac{\partial}{\partial x}\Omega - ij\xi\Omega. \tag{2.5}$$

Now substituting (2.1)–(2.5) into (1.4) and comparing the coefficients of like powers of z in the right- and left-hand sides, we conclude that (1.4) is equivalent to the nonlinear Schrödinger equation

$$R_t = \frac{i}{2}(R_{xx} - 2RR^*R). \tag{2.6}$$

Let us also introduce the matrix

$$T = \frac{1}{\sqrt{2}}\begin{bmatrix} I_m & -I_m \\ I_m & I_m \end{bmatrix}. \tag{2.7}$$

It is readily verified that

$$TjT^* = J, \qquad T^*JT = j, \qquad T^* = T^{-1}. \tag{2.8}$$

Define the matrix $A(x,t)$ as the solution of the problem

$$j\frac{\partial A}{\partial x} = -\xi A, \qquad A(0,t) = T^*. \tag{2.9}$$

A straightforward substitution yields the following assertion.

Proposition 2.1. *If $\mathcal{W}_1(x,t,z)$ is a solution of the system*

$$\frac{\partial \mathcal{W}_1}{\partial x} = G_1(x,t,z)\mathcal{W}_1, \qquad \mathcal{W}_1(0,t,z) = I_n \tag{2.10}$$

and $G_1(x,t,z)$ has the form (2.1), *then*

$$\mathcal{W}(x,t,z) = A^{-1}(x,t)\mathcal{W}_1(x,t,z)T^* \tag{2.11}$$

is a solution of the system

$$\frac{\partial \mathcal{W}}{\partial x} = izJ\mathcal{H}(x,t)\mathcal{W}, \tag{2.12}$$

where

$$J\mathcal{H}(x,t) = A^{-1}(x,t)jA(x,t). \tag{2.13}$$

Thanks to (2.3), we have

$$\xi = -\xi^*. \tag{2.14}$$

Relations (2.9) and (2.13) yield

$$A^*(x,t)jA(x,t) = J, \qquad \mathcal{H}(x,t) = A^*(x,t)A(x,t) > 0. \tag{2.15}$$

The transition to system (2.12) allows us to use Theorem 1.1. In particular, (1.6) holds, and so does the equality

$$V(x,t,z) = A^{-1}(x,t)V_1(x,t,z)A(0,t), \tag{2.16}$$

where $V_1(x,t,z)$ is defined by the system of equations

$$\frac{\partial V_1}{\partial t} = F_1(x,t,z)V_1, \qquad V_1(x,0,z) = I_n. \tag{2.17}$$

Consequently,

$$V^{*-1}(0,t,\bar{z}) = TV_1^{*-1}(0,t,\bar{z})T^*. \tag{2.18}$$

Remark 2.1. The matrix-valued function $V(x,t,z)$ can be found also from the system (1.7) if we take into account that

$$F(x,t,z) = -A^{-1}(x,t)\frac{\partial A}{\partial t} + A^{-1}(x,t)F_1(x,t,z)A(x,t). \tag{2.19}$$

For the given matrix-valued functions $R_0(x)$, $R_1(t)$, $R_2(t)$, let us put

$$R(x,0) = R_0(x), \qquad R(0,t) = R_1(t), \qquad \left.\frac{\partial R}{\partial x}\right|_{x=0} = R_2(t). \tag{2.20}$$

Applying Theorem 1.1 we obtain the following corollary.

Corollary 2.1. *Suppose that the solution of problem* (2.6), (2.20) *obeys the estimates*

$$\left\| \frac{\partial^k R(x,t)}{\partial x^k} \right\| \leq M, \qquad 0 \leq x, t < \infty, \quad k = 0, 1, \tag{2.21}$$

with some constant $M > 0$. *Then the limit* $v(t, z)$ *in* (1.16) *exists, and* (1.22) *holds.*

Proof. Indeed, thanks to (2.15) and Remark 1.2, condition a) of Theorem 1.1 is satisfied. Condition b) of Theorem 1.1 follows from (2.15) and Theorem 2.2. of Chapter 9. Formula (2.2) implies the inequality

$$F_1(x,t,\bar{z})j + jF_1^*(x,t,\bar{z}) > 0, \tag{2.22}$$

which is valid for large $|z|$ in the domain

$$\frac{\pi}{2} + \varepsilon \leq \arg\, z \leq \pi - \varepsilon, \qquad \varepsilon > 0.$$

Therefore, for such values of z the matrix-valued function $V_1(l,t,\bar{z})$ is j-expanding. Hence, by (2.15) and (2.16), $V^*(l,t,\bar{z})$ is J-expanding, and so condition d) of Theorem 1.1 is satisfied. Using again (2.15) and Remark 1.2, we obtain inequality (1.14). Now it finally follows from (1.6) that condition c) is satisfied. The corollary is proved. □

Example 2 (modified Korteweg-de Vries equation). Let $G_1(x,t,z)$ and ξ be defined as in Example 1 by (2.1), (2.3), and let $F_1(x,t,z)$ have the form

$$\begin{aligned} F_1(x,t,z) = &\begin{bmatrix} z^2 I_m + \frac{1}{2}RR^* & 0 \\ 0 & z^2 I_m + \frac{1}{2}R^*R \end{bmatrix} \begin{bmatrix} izI_m & -R \\ -R^* & -izI_m \end{bmatrix} + \\ &+ \frac{1}{4}\begin{bmatrix} R_x R^* - RR_x^* & R_{xx} + 2izR_x \\ R_{xx}^* - 2izR_x^* & R_x^* R - R^* R_x \end{bmatrix}. \end{aligned} \tag{2.23}$$

It is readily verifies that (1.4) is equivalent to the *modified Korteweg-de Vries equation*

$$R_t = -\frac{1}{4}R_{xxx} + \frac{3}{4}(RR^*R_x + R_x R^* R). \tag{2.24}$$

Formulas (2.7)–(2.19) remain valid in this case as well. Let us put

$$R(x,0) = R_0(x), \qquad \left.\frac{\partial^k}{\partial x^k} R(x,t)\right|_{x=0} = R_{k+1}(t), \qquad k = 0,1,2. \tag{2.25}$$

Corollary 2.2. *Suppose that the solution of problem* (2.24), (2.25) *obeys the estimates*

$$\left\| \frac{\partial^k}{\partial x^k} R(x,t) \right\| \leq M, \qquad 0 \leq x, t < \infty, \quad k = 0, 1, 2, \tag{2.26}$$

with some constant $M > 0$. *Then the limit* $v(t, z)$ *in* (1.16) *exists, and* (1.22) *holds.*

The proof of Corollary 2.1 remains valid for Corollary 2.2 as well, since (2.22) holds for large $|z|$ in the domain

$$0 < \varepsilon \leq \arg\, z \leq \frac{\pi}{4} - \varepsilon.$$

Example 3 (hyperbolic sine-Gordon equation). Let $n = 2$ and

$$G_1(x,t,z) = J(izI_2 - \xi), \qquad F_1(x,t,z) = -\frac{1}{z} g(x,t). \tag{2.27}$$

As in [98], let us put

$$\xi = \begin{bmatrix} 0 & \Psi(x,t) \\ -\Psi(x,t) & 0 \end{bmatrix}, \qquad g(x,t) = i \begin{bmatrix} 0 & \exp\varphi(x,t) \\ \exp\{-\varphi(x,t)\} & 0 \end{bmatrix}. \tag{2.28}$$

From (2.27) and (2.28) it follows that

$$\frac{\partial G_1}{\partial t} = -J \frac{\partial \xi}{\partial t}, \qquad \frac{\partial F_1}{\partial x} = -\frac{1}{z} \frac{\partial g}{\partial x}, \tag{2.29}$$

and

$$[G_1, F_1] = -i[J, g] + \frac{1}{z}[J\xi, g]. \tag{2.30}$$

Substituting relations (2.29) and (2.30) in (1.4) and comparing the coefficients of like powers of z in the right- and the left-hand sides we obtain the system

$$\frac{\partial \psi}{\partial t} = 2 \sinh \varphi, \qquad \frac{\partial \varphi}{\partial x} = 2\psi, \tag{2.31}$$

which in turn yields the (hyperbolic sine-Gordon) equation

$$\frac{\partial^2 \varphi}{\partial x \partial t} = 4 \sinh \varphi. \tag{2.32}$$

Now let us introduce the matrix $A(x,t)$ as the solution of the problem

$$J \frac{\partial A}{\partial x} = -\xi A, \qquad A(0,t) = I_2. \tag{2.33}$$

Proposition 2.2. *If the matrix-valued function $\mathcal{W}_1(x,t,z)$ is a solution of the system* (2.10) *with $G_1(x,t,z)$ taken from* (2.27), *then*

$$\mathcal{W}(x,t,z) = A^{-1}(x,t)\mathcal{W}_1(x,t,z) \tag{2.34}$$

is a solution of the system (2.12), *where*

$$J\mathcal{H}(x,t) = A^{-1}(x,t)JA(x,t). \tag{2.35}$$

Since (2.14) holds, from (2.33), (2.35) it follows that

$$A^*(x,t)JA(x,t) = J, \qquad \mathcal{H}(x,t) = A^*(x,t)A(x,t) > 0. \tag{2.36}$$

Set

$$\varphi(x,0) = \varphi_0(x), \qquad \varphi(0,t) = \varphi_1(t). \tag{2.37}$$

The following corollary can be established in the same way as Corollary 2.1.

Corollary 2.3. *Let $\varphi(x,t)$ be a solution of the system* (2.32), (2.37) *such that*

$$|\varphi(x,t)| \le M, \qquad 0 \le x,t < \infty. \tag{2.38}$$

Then the limit $v(t,z)$ in (1.16) *exists, and* (1.22) *holds.*

Example 4 (matrix Korteweg-de Vries equation). Let $u(x,t)$ be an $m \times m$ matrix-valued function, and let $G_1(x,t,z)$ and $F_1(x,t,z)$ be of the respective forms

$$G_1(x,t,z) = \begin{bmatrix} 0 & I_m \\ U - zI_m & 0 \end{bmatrix}, \tag{2.39}$$

and

$$F_1(x,t,z) = \begin{bmatrix} -U_x & 2(U+2zI_m) \\ 2(U+2zI_m)(U-zI_m) - U_{xx} & U_x \end{bmatrix}. \tag{2.40}$$

Then

$$\frac{\partial G_1}{\partial t} = \begin{bmatrix} 0 & 0 \\ U_t & 0 \end{bmatrix}, \qquad \frac{\partial F_1}{\partial x} = \begin{bmatrix} -U_{xx} & 2U_x \\ 2(U_xU + UU_x + U_xz) - U_{xxx} & U_{xx} \end{bmatrix}, \tag{2.41}$$

and

$$[G_1, F_1] = \begin{bmatrix} -U_{xx} & 2U_x \\ -U_xU - UU_x & U_{xx} \end{bmatrix} - z\begin{bmatrix} 0 & 0 \\ -2U_x & 0 \end{bmatrix}. \tag{2.42}$$

Substituting expressions (2.41) and (2.42) in equation (1.4) we obtain the *matrix Korteweg-de Vries equation*

$$U_t + U_{xxx} - 3(U_xU + UU_x) = 0. \tag{2.43}$$

Let us also introduce the following reduction condition:

$$U(x,t) = U^*(x,t). \tag{2.44}$$

Denote

$$P(x,t) = \begin{bmatrix} 0 & I_m \\ U(x,t) & 0 \end{bmatrix}, \qquad T = \frac{1}{\sqrt{2}}\begin{bmatrix} iI_m & I_m \\ iI_m & -I_m \end{bmatrix},$$

$$\frac{\partial A}{\partial x} = PA, \qquad A(0,t) = T.$$

Then the following assertion holds true.

Proposition 2.3. *If the matrix-valued function $\mathcal{W}_1(x,t,z)$ is a solution of the system* (2.10) *with $G_1(x,t,z)$ taken from (2.39), then*

$$\mathcal{W}(x,t,z) = A^{-1}(x,t)\mathcal{W}_1(x,t,z)A(0,t) \tag{2.45}$$

is a solution of the system (2.12), *where*

$$\mathcal{H}(x,t) = A^*(x,t)\begin{bmatrix} I_m & 0 \\ 0 & 0 \end{bmatrix} A(x,t) \geq 0. \tag{2.46}$$

Now let us put

$$U(x,0) = U_0(x), \qquad \left.\frac{\partial^k U}{\partial x^k}\right|_{x=0} = U_k(t), \qquad k = 0,1,2. \tag{2.47}$$

Corollary 2.4. *Suppose that a solution $U(x,t)$ of the mixed problem* (2.43), (2.47) *satisfies the reduction condition* (2.44) *and the estimate*

$$\left\| \frac{\partial^k U(x,t)}{\partial x^k} \right\| \leq M, \qquad 0 \leq x,t < \infty, \quad k = 0,1,2. \tag{2.48}$$

Then the limit $v(t,z)$ in (1.16) *exists and* (1.22) *holds.*

Indeed, the Hamiltonian $\mathcal{H}(x)$ is of a positive type. By Remark 1.2, condition a) of Theorem 1.1 is satisfied. In order to verify condition b), let us introduce the operator L_0 in the space $L^2_m(0,\infty)$ of column vectors, which acts by the rule

$$L_0 Y = -\frac{d}{dx^2}Y + U_0(x)Y, \qquad 0 \leq x < \infty,$$

on the set of compactly supported vector-valued functions $Y(x)$ with absolutely continuous derivative that satisfy the condition $Y(0) = Y'(0) = 0$. The fact that $U_0(x)$ is bounded implies that the operator $\bar{L}_0$ has deficiency indices (m,m) ([57]). It follows from this fact and Proposition 1.2 that condition b) of Theorem 1.1 is fulfilled. Condition c)–d) can be verified as in Corollary 2.1.

12.3. On the unique solvability of the mixed problem

12.3.1. In general terms, a mixed problem for a partial differential equation is said to be *uniquely solvable* if for appropriately given initial and boundary conditions the solution exists and is unique. For linear equations with constant coefficients a criterion for deciding whether a mixed problem is uniquely solvable or not is known [96]. In the nonlinear case the situation is far more complicated. Let us consider the mixed problem generated by the equation (1.4) and the conditions

$$G_1(x,0,z) = g(x,z), \qquad F(0,t,z) = f(t,z), \tag{3.1}$$

where $g(x,z)$ and $f(t,z)$ are given matrix-valued functions. The following proposition gives a necessary condition for the solvability of the mixed problem (1.4), (3.1).

Proposition 3.1. *Suppose the assumptions of Theorem* 1.1 *are satisfied and the matrix-valued function* $v(t_0, z)$ *constructed according to* (1.22) *is not in the Nevanlinna class. Then the mixed problem* (1.4), (3.1) *has no solution in the domain* $0 \le x < \infty$, $0 \le t \le t_0$.

12.3.2. Using Proposition 3.1, we can prove that in a number of cases the initial conditions $G(x, 0, z) = g(x, z)$ are uniquely determined and therefore can be omitted once the boundary conditions $F(0, t, z) = f(t, z)$ are given.

Lemma 3.1. *Let* $n = 2$,

$$\operatorname{Tr} F^*(0, t, \bar{z}) = 0 \tag{3.3}$$

and the assumptions of Theorem 1.1 *be satisfied. Then* (1.2) *holds and*

$$r_{11}(t, z)r_{22}(t, z) - r_{12}(t, z)r_{21}(t, z) = 1. \tag{3.4}$$

Indeed, identity (3.4) follows directly from (1.7), (1.9) and (3.3).

Theorem 3.1. *Under the assumptions of Lemma* 3.1, *suppose that for some* z *the following inequality holds:*

$$F^*(0, t, \bar{z})J + JF^*(0, t, \bar{z}) \ge \delta I_2, \qquad \delta > 0. \tag{3.5}$$

Then

$$v(z) = -i \lim_{t\to\infty} \frac{r_{22}(t, z)}{r_{21}(t, z)} = -i \lim_{t\to\infty} \frac{r_{12}(t, z)}{r_{11}(t, z)}. \tag{3.6}$$

Proof. By relations (1.9) and (3.4), we have

$$V^*(0, t, \bar{z}) = \begin{bmatrix} r_{22}(t, z) & -r_{12}(t, z) \\ -r_{21}(t, z) & r_{11}(t, z) \end{bmatrix}. \tag{3.7}$$

Further, by (1.7) and (3.5),

$$\begin{aligned} \frac{d}{dt}B(t, z) &= V^*(0, t, \bar{z})[F^*(0, t, \bar{z})J + JF(0, t, \bar{z})]V(0, t, \bar{z}) \\ &\ge \delta V^*(0, t, \bar{z})V(0, t, \bar{z}), \end{aligned} \tag{3.8}$$

where we use the notation $B(t, z) = V^*(0, t, \bar{z})JV(0, t, \bar{z})$. This inequality yields

$$B(t, z) \ge J + \delta \int_0^t V^*(0, t, \bar{z})V(0, t, \bar{z})\, dt. \tag{3.9}$$

Comparing the diagonal entries in (3.9), we obtain the inequality

$$-[r_{11}(t, z)\overline{r_{21}(t, z)} + r_{21}(t, z)\overline{r_{11}(t, z)}] \ge \delta \int_0^t \left[|r_{21}(t, z)|^2 + |r_{11}(t, z)|^2\right] dt,$$

whence

$$-[r_{11}(t,z)\overline{r_{21}(t,z)} + r_{21}(t,z)\overline{r_{11}(t,z)}] \to +\infty, \qquad t \to +\infty. \tag{3.10}$$

From (3.4) and (3.10) it follows that

$$\lim_{t\to+\infty}\left[\frac{r_{22}}{r_{21}} - \frac{r_{12}}{r_{11}}\right] = 0. \tag{3.11}$$

Using again relation (3.4), we can recast (1.22) as

$$v(t,z) = i\frac{r_{11}}{r_{21}}\left\{1 - \frac{1}{r_{11}r_{21}}\left[-iv(z) + \frac{r_{22}}{r_{21}}\right]^{-1}\right\}. \tag{3.12}$$

Now suppose (3.6) is false. Then, by (3.12),

$$v(t_k,z) = i\frac{r_{11}}{r_{21}}\left[1 + o\left(\frac{1}{r_{11}r_{21}}\right)\right], \qquad t_k \to +\infty, \tag{3.13}$$

which in conjunction with (3.10) shows that

$$\operatorname{Im} v(t_k,z) = \frac{1}{2|r_{21}|^2}\left[(r_{11}\bar{r}_{21} + \bar{r}_{21}r_{11}) + O(1)\right], \qquad t_k \to +\infty.$$

In view of (3.10), this implies that $\operatorname{Im} v(t_k,z) < 0$ for sufficiently large t_k, which contradicts the fact that $v(t_k,z)$ is in the Nevanlinna class. Relation (3.6) is proved. □

12.3.3. For $n = 2$, in examples 1–4 of Section 12.2 all the assumptions of Theorem 3.1 are satisfied. Hence, (3.6) holds. The value of the solution for $t = 0$ is uniquely determined by $v(z)$. Therefore, the following assertions hold true.

Proposition 3.2. *The scalar nonlinear Schrödinger equation* (2.6) *with the boundary conditions*

$$R(0,t) = R_1(t), \qquad \frac{\partial}{\partial x}R\bigg|_{x=0} = R_2(t) \tag{3.14}$$

has at most one solution in the class of functions satisfying (2.21).

Proposition 3.3. *The scalar modified Korteweg-de Vries equation* (2.24) *with the boundary conditions*

$$R(0,t) = R_1(t), \qquad \frac{\partial}{\partial x}R\bigg|_{x=0} = R_2(t), \qquad \frac{\partial^2}{\partial x^2}R\bigg|_{x=0} = R_3(t) \tag{3.15}$$

has at most one solution in the class of functions satisfying (2.26).

Proposition 3.4. *The hyperbolic sine-Gordon equation* (2.32) *with the boundary conditions*

$$\varphi(0,t) = \varphi_1(t) \tag{3.16}$$

has at most one solution in the class of functions satisfying (2.38).

Proposition 3.5. *The scalar Korteweg-de Vries equation* (2.43) *with the boundary conditions*

$$U(0,t) = U_1(t), \qquad \left.\frac{\partial U}{\partial x}\right|_{x=0} = U_2(t), \qquad \left.\frac{\partial^2 U}{\partial x^2}\right|_{x=0} = U_3(t) \tag{3.17}$$

has at most one solution in the class of real functions satisfying (2.48).

12.4. A hierarchy of nonlinear equations and asymptotic behaviour of Weyl-Titchmarsh functions

Let the matrix-valued functions $G(x,t,z)$ and $F_N(x,t,z)$ have the forms

$$G(x,t,z) = g_1(x,t) + zg_0(x,t)), \tag{4.1}$$

and

$$F_N(x,t,z) = f_N(x,t) + zf_{N-1}(x,t)) + \ldots + z^N f_0(x,t), \qquad 1 \le N < \infty. \tag{4.2}$$

The $2m \times 2m$ matrix-valued functions $g_k(x,t)$ $(k = 0,1)$ and $f_k(x,t)$ $(0 \le k \le N)$ are assumed to be differentiable. Moreover, assume that

$$g_0 = ij, \qquad f_0 = ij, \qquad j = \begin{bmatrix} I_m & 0 \\ 0 & -I_m \end{bmatrix}, \tag{4.3}$$

$$g_1 = \begin{bmatrix} 0 & R_{12}(x,t) \\ R_{21}(x,t) & 0 \end{bmatrix}, \tag{4.4}$$

where $R_{12}(x,t)$, $R_{21}(x,t)$ are $m \times m$ matrices. Now consider the matrix equations

$$\frac{\partial W(x,t,z)}{\partial x} = G(x,t,z)W(x,t,z), \tag{4.5}$$

$$\frac{\partial W(x,t,z)}{\partial t} = F_N(x,t,z)W(x,t,z). \tag{4.6}$$

As is known, for each $N = 1,2,3,\ldots$ the condition of compatibility of the systems (4.5) and (4.6) has the form

$$\frac{\partial G}{\partial t} - \frac{\partial F_N}{\partial x} + [G, F_N] = 0. \tag{4.7}$$

The partial differential equations (4.7) form a *hierarchy.* When $N = 2$ equation (4.7) can be reduced to the well-known *matrix nonlinear Schrödinger equation* and when $N = 3$ it can be reduced to the matrix modified Korteweg-de Vries equation.

Let us assume that the reduction condition

$$R_{12}(x,t) = R_{21}^*(x,t) \tag{4.8}$$

is satisfied. In this case the linear problem (4.5) is selfadjoint, i.e., we can use results of spectral theory. In particular, the Weyl-Titchmarsh $m \times m$ matrix function $v(t,z)$ is defined by the following inequality (see Chapter 9):

$$\int_0^\infty [I_m \quad iv^*(t,z)] U \mathcal{W}^*(x,t,z) \mathcal{W}(x,t,z) U^* \begin{bmatrix} I_m \\ -iv(t,z) \end{bmatrix} dx < \infty,$$

where $\mathcal{W}(x,t,z)$ is a solution of system (4.5) satisfying the condition

$$\mathcal{W}(0,t,z) = I_{2m}$$

and

$$U = \frac{1}{\sqrt{2}} \begin{bmatrix} I_m & -I_m \\ I_m & I_m \end{bmatrix}. \tag{4.9}$$

From Theorem 1.1 we deduce the following assertion.

Theorem 4.1. *Suppose that relations* (4.7) *and* (4.8) *hold and*

$$\left\| \frac{\partial^k R_{12}(x,t)}{\partial x^k} \right\| \leq M, \quad 0 \leq x < \infty, \quad 0 \leq t \leq T, \qquad 0 \leq k \leq N.$$

Then the corresponding system (4.5) *has the unique Weyl-Titchmarsh function* $v_N(t,z)$*, which is defined by means of the equality*

$$\begin{aligned} v_N(t,z) = i\{r_{11}(t,z,N)[-iv(z)] + r_{12}(t,z,N)\} \times \\ \times \{r_{21}(t,z,N)[-iv(z)] + r_{22}(t,z,N)\}^{-1}. \end{aligned} \tag{4.10}$$

Here $v(z) = v(0,z)$ *and the matrix* $r_N(t,z) = \{r_{ij}(t,z,N)\}_{i,j=1}^2$ *is defined by the relation*

$$\frac{dr_N(t,z)}{dt} = -\widetilde{F}_N^*(0,t,\bar{z}) r_N(t,z), \tag{4.11}$$

where

$$r_N(0,z) = I_{2m}, \qquad \widetilde{F}_N(0,t,z) = U F_N(0,t,z) U^*. \tag{4.12}$$

Let us decompose the matrix $\widetilde{F}_N(0,t,z)$ into $m \times m$ blocks:

$$\widetilde{F}_N(0,t,z) = \left\{ \widetilde{F}_{kl}(t,z,N) \right\}_{k,l=1}^2.$$

Differentiating both sides of equality (4.10) with respect to t and using (4.11), we obtain

$$\begin{aligned} \frac{dv_N}{dt} &= -\widetilde{F}^*_{11}(t,\bar{z},N)v_N - i\widetilde{F}^*_{21}(t,\bar{z},N) + \\ &+ v_N\widetilde{F}^*_{22}(t,\bar{z},N) - iv_N\widetilde{F}^*_{12}(t,\bar{z},N)v_N. \end{aligned} \tag{4.13}$$

12.4.2. We shall further assume that in a neighbourhood of $z=\infty$ the function $v_N(t,z)$ admits the expansion

$$v_N(t,z) = \sum_{k=0}^{\infty} \frac{\alpha_k(t,N)}{z^k}, \qquad \alpha_0(t,N) = iI_m. \tag{4.14}$$

Definition 4.1. Let relations (4.7) and (4.8) be fulfilled. We say that the matrix-valued function $R_{12}(x,t)$ *belongs to the regularity class* $\mathfrak{P}_T(N)$ if the Weyl-Titchmarsh function $v_N(t,z)$ of system (4.5) admits the representation (4.14) ($0 \le t \le T$) in a neighbourhood of $z=\infty$.

Remark 4.1. The matrix $R_{12}(x,t)$ is a solution of the nonlinear equation generated by equation (4.7).

Thanks to (4.2) and (4.12), the matrix-valued function $\widetilde{F}^*_{kl}(t,\bar{z},N)$ admits the representation

$$\widetilde{F}^*_{kl}(t,\bar{z},N) = \sum_{s=0}^{N} \Psi_{kl}(t,s) z^{N-s}. \tag{4.15}$$

Substituting this expression in (4.13) and comparing the coefficients of like powers of z we obtain

$$\begin{aligned} \frac{d}{dt}\alpha_k(t,N) &= -\sum_{p+s-N=k} \Psi_{11}(t,s)\alpha_p(t,N) \\ + \sum_{p+s-N=k} \alpha_p(t,N)\Psi_{22}(t,s) &- \sum_{p+l+s-N=k} \alpha_p(t,N)\Psi_{12}(t,s)\alpha_l(t,N), \end{aligned} \tag{4.16}$$

where $p \ge 0$, $l \ge 0$, $0 \le s \le N$, and $k \ge 1$.

Further, it follows from relations (4.15) and (4.16) that

$$\begin{aligned} &\sum_{p+s=m} \Psi_{11}(t,s)\alpha_p(t,N) + \sum_{p+s=m} \alpha_p(t,N)\Psi_{22}(t,s) \\ &- \sum_{p+l+s=m} \alpha_p(t,N)\Psi_{12}(t,s)\alpha_l(t,N) - i\Psi_{21}(t,m) = 0, \end{aligned} \tag{4.17}$$

where $p \ge 0$, $l \ge 0$, and $0 \le s \le m \le N$. By (4.3), (4.9) and (4.14),

$$\begin{bmatrix} \Psi_{11}(t,0) & \Psi_{12}(t,0) \\ \Psi_{21}(t,0) & \Psi_{22}(t,0) \end{bmatrix} = i \begin{bmatrix} 0 & I_m \\ I_m & 0 \end{bmatrix}. \tag{4.18}$$

This allows us to rewrite (4.17) in the form

$$\begin{aligned}2i\alpha_m(t,N) &= \sum_{p+s=m} \Psi_{11}(t,s)\alpha_p(t,N) - \sum_{p+s=m} \alpha_p(t,N)\Psi_{22}(t,s) \\ &+ i \sum_{p+l+s=m} \alpha_p(t,N)\Psi_{12}(t,s)\alpha_l(t,N) + i\Psi_{21}(t,m),\end{aligned} \tag{4.19}$$

where $0 \le p$, $l < m$, and $0 < s \le m \le N$. Since the coefficients $\Psi_{ij}(t,s)$ of the recursion relation (4.17) do not depend on N, the following assertion is valid.

Proposition 4.1. *Let* $N_1 \le N_2$. *Then*

$$\alpha_p(t,N_1) = \alpha_p(t,N_2), \qquad 0 \le p \le N_1.$$

Formulas (4.19) hold when $t = 0$ as well. In this case we obtain the equality

$$\begin{aligned}2i\alpha_m(0) &= \sum_{p+s=m} \Psi_{11}(0,s)\alpha_p(0) - \sum_{p+s=m} \alpha_p(0)\Psi_{22}(0,s) \\ &+i \sum_{p+l+s=m} \alpha_p(0)\Psi_{12}(0,s)\alpha_l(0) + i\Psi_{21}(0,m),\end{aligned} \tag{4.20}$$

where $0 \le p$, $l < m$, and $0 < s \le m \le N$.

It is interesting to note that formula (4.20), referring to the *linear stationary* system

$$\frac{dW(x,0,z)}{dx} = G(x,0,z)W(x,0,z), \tag{4.21}$$

was deduced as a result of the investigation of a hierarchy of *nonlinear nonstationary* equations.

Remark 4.2. From formulas (4.2)–(4.4) and (4.12), (4.19) it follows that

$$\alpha_1(t,N) = -R_{12}^*(0,t), \qquad N \ge 1, \tag{4.22}$$

and

$$\alpha_2(t,N) = -\frac{1}{2}i\left[\frac{\partial}{\partial x} R_{12}^*(x,t)\big|_{x=0} + 2R_{12}^*(0,t)^2\right], \qquad N \ge 2. \tag{4.23}$$

12.4.3. A consequence of relations (4.1), (4.2), and (4,7) is that

$$-\frac{\partial}{\partial x} f_k(x,t) + [g_1, f_k] + [g_0, f_{k+1}] = 0, \qquad 0 \le k < N, \tag{4.24}$$

$$\frac{\partial}{\partial t} g_1 - \frac{\partial}{\partial x} f_N + [g_1, f_N] = 0. \tag{4.25}$$

Notice that relations (4.24) do not depend on N. These relations allow to deduce recursion formulas for constructing $f_k(x,t)$, $k>0$, i.e., for constructing the hierarchy (4.7). Here we give the formulas for $f_k(x,t)$ only when $k=1,2,3$. Namely,

$$f_1(x,t)=\begin{bmatrix} 0 & R_{12}(x,t) \\ R_{21}(x,t) & 0 \end{bmatrix}, \tag{4.26}$$

$$f_2(x,t)=\frac{i}{2}\begin{bmatrix} R_{12}(x,t)R_{21}(x,t) & -\frac{\partial}{\partial x}R_{12}(x,t) \\ \frac{\partial}{\partial x}R_{21}(x,t) & -R_{21}(x,t)R_{12}(x,t) \end{bmatrix}. \tag{4.27}$$

For $k=3$ the formulas are more complicated:

$$f_3(x,t)=\begin{bmatrix} f_{3,11}(x,t) & f_{3,12}(x,t) \\ f_{3,21}(x,t) & f_{3,22}(x,t) \end{bmatrix}, \tag{4.28}$$

where

$$f_{3,12}(x,t)=-\frac{1}{4}\left[\frac{\partial^2}{\partial x^2}R_{12}(x,t)-2R_{12}(x,t)R_{21}(x,t)R_{12}(x,t)\right] \tag{4.29}$$

and

$$f_{3,21}(x,t)=-\frac{1}{4}\left[\frac{\partial^2}{\partial x^2}R_{21}(x,t)-2R_{21}(x,t)R_{12}(x,t)R_{21}(x,t)\right]. \tag{4.30}$$

We omit the formulas defining the diagonal entries $f_{3,11}(x,t)$ and $f_{3,22}(x,t)$ because they are not used. By (4.30), the equations of the hierarchy under consideration have the following form.

For $N=1$:

$$\frac{\partial}{\partial t}R_{12}=\frac{\partial}{\partial x}R_{12}, \qquad \frac{\partial}{\partial t}R_{21}=\frac{\partial}{\partial x}R_{21}. \tag{4.31}$$

For $N=2$:

$$\begin{cases} \dfrac{\partial R_{12}}{\partial t}=-\dfrac{i}{2}\left(\dfrac{\partial^2 R_{12}}{\partial x^2}-2R_{12}R_{21}R_{12}\right), \\[2ex] \dfrac{\partial R_{21}}{\partial t}=\dfrac{i}{2}\left(\dfrac{\partial^2 R_{21}}{\partial x^2}-2R_{21}R_{12}R_{21}\right). \end{cases} \tag{4.32}$$

If one of the reduction conditions

$$R_{12}(x,t)=\pm R_{21}^{*}(x,t) \tag{4.33}$$

is satisfied, then system (4.32) is equivalent to the matrix nonlinear Schrödinger equation

$$\frac{\partial R_{12}}{\partial t}=-\frac{i}{2}\left(\frac{\partial^2 R_{12}}{\partial x^2}\mp 2R_{12}R_{12}^{*}R_{12}\right). \tag{4.34}$$

Finally, if one of the reduction conditions (4.33) is satisfied, then in view of (4.29) the $N=3$ equation of the hierarchy coincides with the matrix modified Korteweg-de Vries equation

$$\frac{\partial R_{12}}{\partial t}=-\frac{1}{4}\left(\frac{\partial^3 R_{12}}{\partial x^3}\mp 3\frac{\partial R_{12}}{\partial x}R_{12}^{*}R_{12}\mp 3R_{12}R_{12}^{*}\frac{\partial R_{12}}{\partial x}\right). \tag{4.35}$$

Comments

Chapter 1

Conjecture. *Assume that the matrix-valued function* $\mathcal{W}(z)$ *has the form*

$$\mathcal{W}(z) = I + \sum_{k=1}^{m}\sum_{i=1}^{l} \frac{D_{k,i}}{(z_k - z)^i} \tag{1}$$

and admits the factorization $\mathcal{W}(z) = \mathcal{W}_2(z)\mathcal{W}_1(z)$, *where* $N_1 = N_2$ *(see Theorem* 3.1). *Then*

$$\delta(\mathcal{W}_1) + \delta(\mathcal{W}_2) - \delta(\mathcal{W}) = \dim\ker(S_{11}). \tag{2}$$

It follows from Theorem 3.1 that this conjecture is true in the special case when one of the sides of equality (2) is equal to zero.

Chapter 2

The problem of decomposing $m \times m$ matrix-valued functions $\mathcal{W}(z)$ into a product of real factors is an important special case of the factorization problem.

A matrix-valued function $\mathcal{W}(z)$ is said to be $j_{\mathfrak{G}}$*-real* if it satisfies the relation

$$j_{\mathfrak{G}}\mathcal{W}(z)j_{\mathfrak{G}} = \mathcal{W}(\bar{z}),$$

where $j_{\mathfrak{G}}$ is an involution operator in the space $\mathfrak{G}$.

Recall that an operator $j_{\mathfrak{G}}$ is called an *involution operator* if it satisfies the following conditions:

1. $j_{\mathfrak{G}}^2 = I_{\mathfrak{G}}$,
2. $j_{\mathfrak{G}}(g_1 + g_2) = j_{\mathfrak{G}}g_1 + j_{\mathfrak{G}}g_2$,
3. $j_{\mathfrak{G}}(\alpha g) = \bar{\alpha} g$,
4. $(j_{\mathfrak{G}}g_1, j_{\mathfrak{G}}g_2) = \overline{(g_1, g_2)}$.

Thus, an involution operator is an analogue of the operation of complex conjugation.

In the paper [82] we have shown how to decompose $j_{\mathfrak{G}}$-real rational $m \times m$ matrix-valued functions $\mathcal{W}(z)$ (see formula (1)) into elementary $j_{\mathfrak{G}}$-real factors:

$$\mathcal{W}(z) = \mathcal{W}_n(z)\mathcal{W}_{n-1}(z)\cdots\mathcal{W}_1(z).$$

The problem of the decomposition into a continual product of entire real 2×2 matrix-valued functions $\mathcal{W}(z)$ satisfying $\det\mathcal{W}(z) \neq 0$ was investigated in a very detailed manner in a number of works by L. de Branges (see [13]).

The decomposition of $j_{\mathfrak{G}}$-real entire $m \times m$ matrix-valued functions $\mathcal{W}(z)$ satisfying $\det\mathcal{W}(z) \neq 0$ into a continual product of elementary $j_{\mathfrak{G}}$-real factors is given in the book by I. C. Gohberg and M. G. Kreĭn ([34], Ch. 5).

Chapter 3

1. From the results of Section 3.2 one can derive in a straightforward manner the following assertion (see [80], [81]).

Theorem 1. *If the characteristic operator function $W(z)$ of the operator A satisfies the condition*

$$\|W(z)\| \leq M, \qquad z \neq \bar{z},$$

for some constant $M > 0$, then the simple part of A is linearly similar to a selfadjoint operator.

Let also note that the results of Section 3.2 proved useful in constructing a scattering theory for non-selfadjoint operators [80], [81].

2. The papers [28] and [79] are devoted to the question of uniqueness of representation (2.30). The most complete results have been obtained for the case $J = I$.

Chapter 4

1. The advantage of the integral form (1.10) over the differential form (1.14) is that it allows one to investigate more general problems.

2. Multiplicative representations (2.6) (Theorem 2.1) and (3.4) (Theorem 3.2) were obtained in the work by I. Gohberg and M. G. Kreĭn [32] by a different method and under some other conditions.

3. Theorem 2.4 was obtained by using the method of operator identities and by following V. P. Potapov's scheme of investigating interpolation problems (see [36], [38]). A complete proof of Theorem 2.4 is given in [88] (Ch. 1).

4. The spectral matrix function $\tau(u)$ was introduced by resorting to the operator V (see (4.1.18)). Let us also introduce the notion of a *quasi-spectral matrix function* (see [74]).

Definition 1. A monotonically increasing matrix-valued function $\tau(u)$ is called a *a quasi-spectral matrix function* of the system (4.1.5), (4.1.13) if the operator V maps the subspace $L^2(B_1) \ominus \ker V$ isometrically into $L^2(\tau)$.

Further we shall assume that the following condition is satisfied.

Condition 1. If $c(z)g = 0$ then necessarily $g = 0$ (here $c(z)$ is defined by the formula (4.2.27)).

Proposition 1. *If* $\ker \Phi_2 = 0$ *and the operator S is positive and invertible then Condition* 1 *is satisfied for the system* (4.1.10), (4.2.1) *generated by the corresponding S-node.*

This assertion is a direct consequence of the equality

$$\frac{i[c(z)d^*(z) + d(z)c^*(z)]}{z - \bar{z}} = \Phi_2^*(I - zA^*)^{-1}S^{-1}(I - \bar{z}A)^{-1}\Phi_2.$$

Proposition 2. *Let $\tau(u)$ be a quasi-spectral matrix function of the system* (4.1.10), (4.2.1) *and let Condition* 1 *be satisfied. Then*

$$\int_{-\infty}^{\infty} \frac{d\tau(u)}{1 + u^2} < \infty. \tag{3}$$

Remark 1. It follows from Propositions 1 and 2 that in Theorem 4.2.6 the requirement that the integral (3) converges can be omitted.

5. As it was proved by M. S. Livšic [58], the operator A_{M} (see (4.3.8)) is a triangular model of an operator whose spectrum is concentrated in zero and $\mathcal{W}(l, z)$ is the characteristic matrix function of this operator.

3. Chapter 3 contains a method of constructing a canonical system from a known monodromy matrix $W(z) = \mathcal{W}(l, z)$ (see Chapter 3, Theorem 1.1.) The problem of existence, i.e., the description of the class of the monodromy matrices $W(z)$ generating canonical systems is completely solved in the well-known work by V. P. Potapov [65].

Chapter 9

The results of Sections 3 and 4 were obtained in the papers [85] and [95], respectively.

Chapter 10

1. For the case $m = 1$, $d_1 = 1$ formula (6.15) is given in M. G. Kreĭn's paper [48] without a proof.

2. For the scalar case ($m = 1$) M. G. Kreĭn proved the following assertion [48].

Proposition 1. *The spectral function* $\tau(u)$ *of the system* (0.8) *is orthogonal if and only if*

$$\int_{-\infty}^{\infty} \frac{\ln \tau'(u)}{1+u^2}\, du = -\infty.$$

For the general case ($m \geq 1$) we make the the following

Conjecture. *The spectral matrix function* $\tau(u)$ *of the system* (1.17) *is orthogonal if*

$$\int_{-\infty}^{\infty} \frac{\ln \|\tau'(u)\|}{1+u^2}\, du = -\infty.$$

3. In the same work by M. G. Kreĭn [48] one finds also the following assertion.

Proposition 2. *In the case* $m = 1$ *the condition*

$$\int_{-\infty}^{\infty} \frac{\ln \tau'(u)}{1+u^2}\, du > -\infty \tag{4}$$

is equivalent to conditions 1)–4) *of Theorem* 1.1.

This important proposition was not proved in [48]. Here we formulate a more general assertion (for the case $m \geq 1$), which has been proved by us.

Proposition 3. *The condition*

$$\int_{-\infty}^{\infty} \frac{\ln \det \tau'(u)}{1+u^2}\, du > -\infty \tag{5}$$

is equivalent to conditions 1)–4) *of Theorem* 1.1.

Chapter 11

In their paper [40], I. S. Kac and M. G. Kreĭn introduced the important notion of a *dual string.* Here we give a generalization of this notion for the matrix string.

Let $\alpha(x)$ and $\rho(x)$ be $m \times m$ matrix-valued functions such that

$$\|\alpha^{\pm 1}(x)\| \leq M, \qquad \|\rho^{\pm 1}(x)\| \leq M.$$

Consider the operators B and C in $L^2_m(0,l)$ acting according to the rules

$$Bf = \alpha(x) \int_0^x \rho(t) f(t)\, dt \tag{6}$$

and

$$Cf = -\rho^*(x) \int_0^x \alpha^*(t) f(t)\, dt. \tag{7}$$

Then one can readily verify that

$$(B^* - C)f = \rho^*(x) \int_0^l \alpha^*(t) f(t)\, dt. \tag{8}$$

Equality (8) can be written in the form

$$(B^* - C) = \Pi_2 \Pi_1^*, \tag{9}$$

where the operators Π_1 and Π_2 act on constant $m \times 1$ column vectors g by the rules

$$\Pi_1 g = \alpha(x) g, \qquad \Pi_2 g = \rho^*(x) g. \tag{10}$$

Now introduce the operators

$$A = CB, \qquad \Phi_1 = B^* \Pi_1, \qquad \Phi_2 = i\Pi_2, \tag{11}$$

$$\widetilde{A} = BC, \qquad \widetilde{\Phi}_1 = \Pi_1, \qquad \widetilde{\Phi}_2 = iB\Pi_2. \tag{12}$$

From (9), (11), and (12) we obtain the operator identities

$$A - A^* = i(\Phi_1 \Phi_2^* + \Phi_2 \Phi_1^*) \tag{13}$$

and

$$\widetilde{A} - \widetilde{A}^* = i(\widetilde{\Phi}_1 \widetilde{\Phi}_2^* + \widetilde{\Phi}_2 \widetilde{\Phi}_1^*). \tag{14}$$

We will say that identity (14) is *dual* to identity (13) if relations (9), (11), and (12) hold. With the identities (13) and (14) one associates (see Chapter 4) canonical systems with the Hamiltonians

$$\mathcal{H}(x) = \begin{bmatrix} Q(x) \\ P(x) \end{bmatrix} [Q^*(x) \;\; P^*(x)], \tag{15}$$

and

$$\widetilde{\mathcal{H}}(x)=\begin{bmatrix}\widetilde{Q}(x)\\ \widetilde{P}(x)\end{bmatrix}\begin{bmatrix}\widetilde{Q}^*(x) & \widetilde{P}^*(x)\end{bmatrix}, \tag{16}$$

where

$$Q(x)=\rho^*(x)\int_x^l \alpha^*(t)\alpha(t)\,dt, \qquad P^*(x)=i\rho(x),$$

and

$$\widetilde{Q}^*(x)=\alpha(x), \qquad \widetilde{P}^*(x)=i\alpha(x)\int_0^x \rho(t)\rho^*(t)\,dt.$$

Using formulas (1.15)–(1.19) we obtain the following assertion.

Identities (13) *and* (14) *generate the matrix string equations*

$$-\rho^{-1}(x)\frac{d}{dx}\left\{\alpha^{-1}(x)\alpha^{*-1}(x)\frac{d}{dx}\left[\rho^{*-1}(x)Y(x,z)\right]\right\}=zY(x,z), \tag{17}$$

and

$$-\alpha^{*-1}(x)\frac{d}{dx}\left\{\rho^{*-1}(x)\rho^{-1}(x)\frac{d}{dx}\left[\alpha^{-1}(x)Y(x,z)\right]\right\}=zY(x,z), \tag{18}$$

respectively.

We say that equation (18) is *dual* to equation (17).

Our definition of the dual string coincides with the definition given in the paper [40] when $m=1$ and $\alpha(x)\equiv 1$.

Chapter 12

1. In the papers [87] and [89] we considered the case when $m=1$, $N=2,3$ (see Section 4) and

$$v(0,z)=i+\sum_{k=1}^{n}\frac{\beta_k(0)}{z-\alpha_k(0)},$$

where

$$\operatorname{Im}\alpha_k(0)<0, \qquad \beta_k(0)\neq 0, \qquad \alpha_k(0)\neq\alpha_l(0), \qquad k\neq l.$$

In this case there exists one, and only one solution of equation (4.13) such that

$$v_N(t,z)=i+\sum_{k=1}^{n}\frac{\beta_k(t)}{z-\alpha_k(t)}, \qquad 0\le t\le T,$$

where

$$\operatorname{Im}\alpha_k(t)<0, \qquad \beta_k(t)\neq 0, \qquad \alpha_k(t)\neq\alpha_l(t), \qquad k\neq l.$$

The corresponding solutions $R_{12}(x,t)$ of the system (4.7), (4.8) belong to the regularity class $\mathcal{P}_T$ and are written in explicit form. These solutions have n points of discontinuity $x_1(t), x_2(t), \dots, x_n(t)$ where the solutions become infinite. The motion of these discontinuity points is very similar to the motion of the humps of n-soliton solutions. The dynamics of the motion of the points $x_1(t), x_2(t), \dots, x_n(t)$ turns out to be Hamiltonian and may be interpreted as the motion of n interacting particles [87], [88].

2. In the paper [92] the method of the inverse spectral problem is applied to the investigation of integrable nonlinear equations on the line $-\infty < x < \infty$.

3. Discrete nonlinear equations (Toda chain and others) were investigated by the method of the inverse spectral problem in work by Yu. M. Berezanskiĭ and his students (see, e.g., [11] and [24]) and in the author's papers [86], [88].

References

[1] Ablowitz, M. J., and Segur, H. *Solitons and the Inverse Scattering Transform,* SIAM Studies in Applied Mathematics, 4, Society for Industrial and Applied Mathematics, Philadelphia, Pa., 1981. [MR **84a:35251**]

[2] Ahiezer, N. I., and Glazman, I. M., *Theory of Linear Operators in Hilbert Space,* Vols. I and II, Translated from the third Russian edition, Monographs and Studies in Mathematics, 9 and 10, Pitman (Advanced Publishing Program), Boston, Mass.-London, 1981. [MR **83i:47001a,b**]

[3] Alpay, D., and Gohberg, I., *Inverse spectral problem for difference operators with rational scattering matrix functions,* Integral Equations and Operator Theory, **20**, no. 2, 125–170 (1994). [MR **95m:47053**]

[4] Alpay, D., Gohberg, I., and Sakhnovich, L., *Inverse scattering problem for continuous transmission lines with rational reflection coefficient function,* in: Recent developments in operator theory and its applications (Winnipeg, MB, 1994), 1–16, Oper. Theory Adv. Appl., 87, Birkhäuser, Basel, 1996. [MR **97f:34072**]

[5] Ambartsumyan, V. A., *Scientific Works,* Vol. 1, "Nauka", Erevan (Russian), 1960.

[6] Atkinson, F. V., *Discrete and Continuous Boundary Problems,* Mathematics in Science and Engineering, Vol. 8, Academic Press, New York-London 1964. [MR **31** #416]

[7] Bart, H., Gohberg, I., and Kaashoek, M. A., *Minimal Factorization of Matrix and Operator Functions,* Oper. Theory Adv. Appl., 1. Birkhäuser, Basel-Boston, Mass., 1979. [MR **81a:47001**]

[8] Bart, H., Gohberg, I., and Kaashoek, M. A., *Convolution equations and linear systems,* Integral Equations Operator Theory **5**, no. 3, 283–340 (1982), and *Errata,* **5**, no. 6, 902 (1982). [MR **84k:45008a,b**]

[9] Bart, H., Gohberg, I., and Kaashoek, M. A., *Invariants for Wiener-Hopf equivalence of analytic operator functions,* in: Constructive methods of Wiener-Hopf factorization, Oper. Theory Adv. Appl., 21, 317–356, Birkhäuser, Basel, 1986.

[10] Bellman, R., *Stability Theory of Differential Equations,* McGraw-Hill Book Company, Inc., New York-Toronto-London, 1953. [MR **15,794b**]

[11] Berezanskiĭ, Yu. M., *Integration of nonlinear difference equations by the method of the inverse spectral problem,* Dokl. Akad. Nauk SSSR **281** (1985), no. 1, 16–19. [MR **86h:39002**]; English transl., Soviet Math. Dokl., **31** (1985), 264–267 [Zbl 595.39003]

[12] Borg, G., *Eine Umkehrung der Sturm-Liouvilleschen Eigenwertaufgabe. Bestimmung der Differentialgleichung durch die Eigenwerte,* Acta Math. **78** (1946), pp. 1–96. [MR **7,382d**]

[13] de Branges, L. *Hilbert Spaces of Entire Functions,* Prentice Hall, Inc., Englewood Cliffs, N.J., 1968. [MR **37 #4590**]

[14] Brodskiĭ, M. S., *Triangular and Jordan Representations of Linear Operators,* "Nauka", Moscow, 1969. [MR **41 #4283**]; English transl., Translations of Mathematical Monographs, Vol. 32. American Mathematical Society, Providence, R.I., 1971. [MR **48 #904**]

[15] Bullough, R. K., and Caudrey, P. J. (eds.), *Solitons,* Topics in Current Physics, 17, Springer-Verlag, Berlin-New York, 1980. [MR **82m:35001**]

[16] Coddington, E. A., and Levinson, N., *Theory of Ordinary Differential Equations,* McGraw-Hill Book Company, Inc., New York-Toronto-London, 1955. [MR **16,1022b**]

[17] Daletskiĭ, Yu. L., and Kreĭn, M. G., *Stability of Solutions of Differential Equations in Banach Space,* "Nauka", Moscow, 1970. [MR **50 #5125**]; English transl., Translations of Mathematical Monographs, Vol. 43. American Mathematical Society, Providence, R.I., 1974. [MR **50 #5126**]

[18] Dym, H., and Iacob A., *Positive definite extensions, canonical equations and inverse problems,* in: Topics in operator theory systems and networks (Rehovot, 1983), 141–240, Oper. Theory Adv. Appl., 12, Birkhäuser, Basel-Boston, Mass., 1984. [MR **86g:34025**]

[19] Dym, H., and Kravitsky, N., *On the inverse problem for the string equation,* Integral Equations Operator Theory **1** (1978), no. 2, 270–277. [MR **80c:34017**]

[20] Dym, H., and McKean, H. P., *Gaussian Processes, Function Theory and the Inverse Spectral Problem,* Probability and Mathematical Statistics, Vol. 31. Academic Press [Harcourt Brace Jovanovich, Publishers], New York-London, 1976. [MR **56 #6829**]

[21] Efimov, A. V., and Potapov, V. P., *J-expanding matrix-valued functions and their role in the analytic theory of electrical circuits,* Uspekhi Mat. Nauk **28** (1973), no.1(169), 65-130. [MR **52 #15090**]; English transl., Russian Math. Surv., **28**, no. 1, 69–140 (1973). [Zbl 285.94009]

[22] Faddeev L. D., and Takhtadzhyan L. A., *Hamiltonian Approach in the Theory of Solitons,* "Nauka", Moscow, 1986; English transl., *Hamiltonian Methods in the Theory of Solitons,* Springer-Verlag, Berlin-New York, 1987. [MR **89m:58102, 58103**]

[23] Gardner, C. S., Greene, J. M., Kruskal, M. D., and Miura, R. M., *Method for solving the Korteweg-de Vries equation,* Phys. Rev. Lett. **19** (1967), 1095–1097.

[24] Gekhtman, M. I., *Nonabelian nonlinear lattice equations of finite interval,* J. Phys. A., **26** (1993), 6303–6317. [MR **95b:39009**]

[25] Gel'fand, I. M., and Levitan, B. M., *On the determination of a differential equation from its spectral function,* Izv. Akad. Nauk SSSR, Ser. Math., **15** (1951), 309–360; English transl., American Mathematical Society Translations (2) **1** (1955), 253–304. [MR **13,558f, 17,489c**]

[26] Gel'fand, I. M., and Vilenkin, N. Ya., *Generalized functions*, Vol 4, *Some applications of harmonic analysis. Rigged Hilbert spaces,* Gosudarstv. Izdat. Fiz.-Mat. Lit., Moscow, 1961. [MR **26** #4173]; English transl., Academic Press, New York, 1964 . [MR **30** #4152]

[27] Gesztesy, F., and Simon, B., *Uniqueness theorems in inverse spectral theory for one-dimensional Schrödinger Operators,* Trans. Amer. Math. Soc. **348** (1996), no. 1, 349–373. [MR **96e:34030**]

[28] Ginzburg, Yu. P., *Reconstruction of a multiplicative integral from its modulus,* Teor. Funktsiĭ Funktsional. Anal. i Prilozhen., No. 41 (1984), 135–143 (Russian) [MR **85k:47063**] (see also Ginzburg, Yu. P. *On recovering a multiplicative integral from its modulus*, in: Topics in interpolation theory (Leipzig, 1994), 185–193, Oper. Theory Adv. Appl., 95, Birkhäuser, Basel, 1997.

[29] Gohberg, I., Goldberg, S., and Kaashoek, M. A., *Classes of Linear Operators,* Vol. I. Oper. Theory Adv. Appl, 49. Birkhäuser Verlag, Basel, 1990. [MR **93d:47002**]

[30] Gohberg, I, Goldberg, S., and Kaashoek, M. A., *Classes of Linear Operators,* Vol. II. Oper. Theory Adv. Appl., 63. Birkhäuser Verlag, Basel, 1993. [MR **95a:47001**]

[31] Gohberg, I., Kaashoek, M. A., Lerer, L., and Rodman, L., *Common multiples and common divisors of matrix polynomials. I. Spectral method,* Indiana Univ. Math. J. 30 (1981), no. 3, 321–356. [MR **82j:15019**]

[32] Gohberg, I. C., and Kreĭn, M. G., *On a description of contraction operators similar to unitary ones,* Funkcional. Anal. i Priložen., **1** (1967), no.1, 38–60. [MR **35** #4761]; English transl., Funct. Anal. Appl. (1967), no.1, 33-52. [Zbl. 161.11601]

[33] Gohberg, I. C., and Kreĭn, M. G., *Introduction to the Theory of Linear Non-selfadjoint Operators in Hilbert Space,* "Nauka", Moscow, 1965. [MR **36** #3137]; English transl., Translations of Mathematical Monographs, Vol. 18, American Mathematical Society, Providence, R.I., 1969. [MR **39** #7447]

[34] Gohberg, I. C., and Kreĭn, M. G., *Theory of Volterra Operators in Hilbert Space and its Applications,* "Nauka", Moscow, 1967. [MR **36** #2007]; English transl., *Theory and Applications of Volterra Operators in Hilbert Space,* Translations of Mathematical Monographs, Vol. 24, American Mathematical Society, Providence, R.I. 1970. [MR **41** #9041]

[35] Helton, J. W., and Ball, J. A., *The cascade decomposition of a given system vs. the linear fractional decompositions of its transfer function,* Integral Equations Operator Theory, **5** (1982), 341–385. [MR **84h:93018**]

[36] Ivanchenko, T. S., and Sakhnovich, L. A., *Operator identities in the theory of interpolation problems,* Izv. Akad. Nauk Armyan. SSR Ser. Mat., **22** (1987), no. 3, 298–308. [MR **89a:47023**]; English transl., Sov. J. Contemp. Math. Anal., Arm. Acad. Sci. **22** (1987), no.3, 84–94. [Zbl 633.30032]

[37] Ivanchenko, T. S., and Sakhnovich, L. A., *An operator approach to V.P. Potapov's scheme for the investigation of interpolation problems,* Ukrain. Mat. Zh., **39** (1987), no. 5, 573–578 (Russian). [MR **89d:47085**]

[38] Ivanchenko, T. S., and Sakhnovich, L. A., *An operator approach to the Potapov scheme for the solution of interpolation problems,* in: Matrix and operator valued functions, 48–86, Oper. Theory Adv. Appl., 72, Birkhäuser, Basel, 1994. [MR **95m:47021**]

[39] Jakubovich, V. A., *The small parameter method for canonical systems with periodic coefficients,* Prikl. Mat. Mekh. **23** (1959), 15–34 (Russian); English transl., J. Appl. Math. Mech., **23** (1959), 17–43. [MR **21** #5774]

[40] Kac, I. S., and Kreĭn, M. G., *On the spectral functions of the string,* in: Nine papers in analysis, American Mathematical Society Translations (2) **103**, 19–102, 1974. [MR **48** #6969]

[41] Kailath, T., *Linear Systems,* Prentice-Hall Information and System Sciences Series, Prentice-Hall, Inc., Englewood Cliffs, N.J., 1980. [MR **82a:93001**]

[42] Kalman, R. E., *Mathematical description of linear dynamical systems,* J. SIAM Control Ser. A **1** (1963), 152–192. [MR **27** #2147]

[43] Kalman, R. E., Falb, P. L., and Arbib, M. A., *Topics in Mathematical System Theory,* McGraw-Hill Book Co, New York-Toronto, 1969. [MR **40** #8465]

[44] Kostenko, N. M., *Necessary and sufficient conditions for factorization of a rational operator-valued function.* Funktsional. Anal. i Prilozhen., **12** (1978), no. 4, 87–88. [MR **80c:47020**]; English transl., Funct. Anal. Appl. **12** (1979), 315-317. [Zbl. 427.47011]

[45] Kovalishina, I. V., and Potapov, V. P., *An indefinite metric in the Nevanlinna-Pick problem,* Akad. Nauk Armjan. SSR Dokl., **59** (1974), no. 1, 17–22. [MR **53** #13577]; English transl., in: Seven papers translated from the Russian, American Mathematical Society Translations, Series 2. 138 [MR **89f:00030**]

[46] Kovalishina I. V., and Potapov, V. P., *Integral representation of hermitian positive functions,* Translated from the Russian by T. Ando, Hokkaido University, Sapporo, 1982. [MR **85e:47019**]

[47] Kovalishina, I. V., *Analytic theory of a class of interpolation problems,* Izv. Akad. Nauk SSSR Ser. Mat., **47** (1983), no. 3, 455–497. [MR **84i:30043**]; English transl., Math. USSR, Izv. **22** (1984), 419-463. [Zbl 549.30026]

[48] Kreĭn, M. G., *Continuous analogues of propositions on polynomials orthogonal on the unit circle,* Dokl. Akad. Nauk SSSR **105** (1955), 637–640 (Russian). [MR **18,291b**]

[49] Kreĭn, M. G., *Analytic problems and results of the theory of linear operators in Hilbert space,* Proc. Internat. Congr. Math. (Moscow, 1966), pp 189–216, "Mir", Moscow, 1968 (Russian). [MR **38** #3733]

[50] Kreĭn, M. G., *Topics in Differential and Integral Equations and Operator Theory,* Edited by I. Gohberg, Oper. Theory Adv. Appl., 7. Birkhäuser Verlag, Basel-Boston, Mass., 1983. [MR **86m:00014**]

[51] Kreĭn, M. G., and Langer, H., *Some propositions on analytic matrix functions related to the theory of operators in the space* Π_κ, Acta Sci. Math. (Szeged) **43** (1981), no. 1–2, 181–205. [MR **82i:47053**]

[52] Kreĭn, M. G., and Lyubarskiĭ, G. Ya., *Analytic properties of the multipliers of periodic canonical differential systems of positive type,* Izv. Akad. Nauk SSSR Ser. Mat., **26** (1962), 549–572 [MR **26** #400]; English transl., American Mathematical Society Translations, II. **89** (1970), 1–28. [Zbl 208.12001]

[53] Kwakernaak, H., and Sivan, R., *Linear Optimal Control Systems,* Wiley-Interscience, New York, 1972. [MR **53** #10394]

[54] Lax, P. D., *Integrals of nonlinear equations of evolution and solitary waves,* Comm. Pure and Appl. Math., **21** (1968), 467–490. [MR **8** #3620]

[55] Leibenzon, Z. L., *The relation between the inverse problem and the completeness of eigenfunctions,* Dokl. Akad. Nauk SSSR **145** (1962), 519–522. [MR **26** #387]; English transl., Soviet Math. Dokl., **3** (1962), 1045–1048.

[56] Levitan, B. M., *Inverse Sturm-Liouville Problems,* "Nauka", Moscow, 1984. [MR **86d:34002**]; English transl., VSP, Zeist, 1987. [MR **89b:34001**]

[57] Levitan, B. M., and Sargsyan, I. S., *Introduction to the Spectral Theory. Selfadjoint Ordinary Differential Operators,* "Nauka", Moscow, 1970 . [MR **45** #8911]; English transl., Translations of Mathematical Monographs, Vol. 39. American Mathematical Society, Providence, R.I., 1975. [MR **51** #6026]

[58] Livšic, M. S., *Operators, Oscillations, Waves. Open Systems,* "Nauka", Moscow, 1966. [MR **38** #1922]; English transl., Translations of Mathematical Monographs, Vol. 34. American Mathematical Society, Providence, R.I., 1973 [MR **49** #12116]

[59] Marchenko, V. A., *Sturm-Liouville Operators and their Applications,* "Naukova Dumka", Kiev, 1977. [MR **58** #1317]; English transl., Oper. Theory Adv. Appl., 22, Birkhäuser Verlag, Basel, 1986. [MR **88f:34034**]

[60] Masani, P., *Recent trends in multivariate prediction theory,* in: Multivariate analysis (Proc. Internat. Sympos., Dayton, Ohio, 1965), pp. 351–382, Academic Press, New York, 1966. [MR **35** #5079]

[61] Melik-Adamyan, F. E., *A class of canonical differential operators,* Izv. Akad. Nauk Armyan. SSR Ser. Mat., **24** (1989), no. 6, 570–592; English transl., Soviet J. Contemporary Math. Anal., **24** (1989), no. 6, 48–69. [MR **92a:47056**]

[62] Zakharov, V. E., Manakov, S. V., Novikov, S. P., and Pitaevskiĭ, L. P., *Theory of solitons. The method of the inverse problem,* "Nauka", Moscow, 1980. [MR **81g:35112**]; English transl., Contemporary Soviet Mathematics, Consultants Bureau [Plenum], New York, 1984. [MR **86k:35142**]

[63] Orlov, S. A., *Nested matrix discs that depend analytically on a parameter, and theorems on the invariance of the ranks of the radii of the limit matrix discs,*

Izv. Akad. Nauk SSSR Ser. Mat. **40** (1976), no. 3, 593–644. [MR **54** #13625]; English transl., Math. USSR, Izv. **10** (1976), 565-613. [Zbl 374.34015]

[64] Pólya, G., and Szegö, G., *Problems and theorems in analysis. Vol. II. Theory of functions, zeros, polynomials, determinants, number theory, geometry,* Revised and enlarged translation of the fourth German edition. Springer Study Edition. Springer-Verlag, New York-Heidelberg, 1976. [MR **53** #2]

[65] Potapov, V. P. *The multiplicative structure of J-contractive matrix functions,* (Russian) Trudy Moskov. Mat. Obšč. **4** (1955), 125–236. [MR **17,958f**]; English transl., American Mathematical Society Translations, **15** (1960), 131–243. [MR **22** #5733]

[66] Potapov, V. P., *Collected Papers of V. P. Potapov,* Translated from the Russian and edited by T. Ando, Hokkaido University, Research Institute of Applied Electricity, Division of Applied Mathematics, Sapporo, 1982. [MR **84a:01061**]

[67] Pozin, S. M., and Sakhnovich, L. A., *On one extreme problem in the theory of approximation,* Integral Equations Operator Theory **21** (1995), no. 4, 484–497. [MR **96h:41028**]

[68] Putnam, C. R., *Commutators, perturbations and unitary spectra,* Acta Math., **106** (1961), 215–232. [MR **24** #A3514]

[69] Putnam, C. R., *Commutation Properties of Hilbert Space Operators and Related Topics,* Ergebnisse der Mathematik und ihrer Grenzgebiete, Band 36, Springer-Verlag, New York, 1967. [MR **36** #707]

[70] Riesz, F., and Sz.-Nagy, B., Leçons d'analyse fonctionnelle, Académie des Sciences de Hongrie, Akadémiai Kiadó, Budapest, 1952. [MR **14,286d**]; English transl., *Functional Analysis,* Frederick Ungar Publishing Co., New York, 1955. [MR **17,175i**]

[71] Rosenblum, M., *On the operator equation $BX - XA = Q$,* Duke Math. J. **23** (1956), 263–269. [MR **18,54d**]

[72] Rozanov, Y. A., *Stationary random processes,* Translated from the Russian, Holden-Day, Inc., San Francisco, 1967. [MR **35** #4985]

[73] Sakhnovich, A. L., *A nonlinear Schrödinger equation on the semi-axis and a related inverse problem,* Ukrain. Mat. Zh., **42** (1990), no. 3, 356–363. [MR **91c:34118**]; English transl., Ukr. Math. J. **42** (1990), no.3, 316-323. [Zbl 711.35132]

[74] Sakhnovich, A. L., *Spectral functions of a canonical system of order $2n$,* Mat. Sb. **181** (1990), no. 11, 1510–1524. [MR **92c:47056**]; English transl., Math. USSR Sbornik, **71** (1992), no. 2, 355–369. [Zbl 776.34066]

[75] Sakhnovich, A. L., *Exact solutions of nonlinear equations and the method of operator identities,* Linear Algebra Appl., **182** (1993), 109–126. [MR **94e:35044**]

[76] Sakhnovich, A. L., *Dressing procedure for solutions of nonlinear equations and the method of operator identities,* Inverse Problems **10** (1994), no. 3, 699–710. [MR **95k:35175**]

[77] Sakhnovich, A. L., *Iterated Bäcklund-Darboux transformation and transfer matrix-function (nonisospectral case),* Chaos Solitons Fractals **7** (1996), no. 8, 1251–1259. [MR **97j:58137**]

[78] Sakhnovich, A. L., *Iterated Bäcklund-Darboux transform for canonical systems,* J. Funct. Anal., **144** (1997), no. 2, 359–370. [MR **98i:34123**]

[79] Sakhnovich, L. A., *Dissipative operators with absolutely continuous spectrum,* Trudy Moskov. Mat. Obšč., **19** (1968), 211–270. [MR **40** #3350]; English transl., Trans. Mosc. Math. Soc., **19** (1968), 233–297.

[80] Sakhnovich, L. A., *Operators similar to unitary operators with absolutely continuous spectrum,* (Russian) Funkcional. Anal. i Priložen. **2** (1968), no. 1, 51–63. [MR **38** #569]; English transl., Functional. Anal. and Appl., **2** (1968), no. 1, 48–60.

[81] Sakhnovich, L. A., *The similarity of linear operators,* Dokl. Akad. Nauk SSSR **200** (1971), 541–544. [MR **45** #7510]; English transl., Sov. Math. Dokl., **12** (1971), 1445–1449. [Zbl 232.47004]

[82] Sakhnovich, L. A., *The factorization of an operator-valued tranfer function,* Dokl. Akad. Nauk SSSR **226** (1976), no. 4, 781–784. [MR **53** #11427]; English transl., Sov. Math. Dokl.,**17** (1976), 203–207. [Zbl 342.93022]

[83] Sakhnovich, L. A., *Inverse spectral problems for systems of equations,* Dokl. Akad. Nauk SSSR **284** (1985), no. 1, 57–61. [MR **87d:47017**]; English transl., Sov. Math., Dokl., **32** (1985), 389–392. [Zbl 617.47012]

[84] Sakhnovich, L.A., *Factorization problems and operator identities,* Uspekhi Mat. Nauk **41** (1986), no. 1, 4–55. [MR **87k:47041**]; English transl., Russian Math. Surv., **41** (1986), no.1, 1–64. [Zbl 613.47017]

[85] Sakhnovich, L. A., *Spectral problems for systems of equations on the axis,* Dokl. Akad. Nauk SSSR **296** (1987), no. 5, 1052–1056. [MR **89e:34049**] ; English transl., Soviet Math. Dokl., **36** (1988), no. 2, 356–360. [Zbl 666.34025]

[86] Sakhnovich, L. A., *On a semi-infinite Toda chain,* Teoret. Mat. Fiz., **81** (1989), no. 1, 12–23; English transl., Theoret. and Math. Phys., **81** (1989), no. 1, 1018–1026. [MR **91i:34096**]

[87] Sakhnovich, L. A., *Integrable nonlinear equations on the semi-axis,* Ukrain. Mat. Zh., **43** (1991), no. 11, 1578–1584. [MR **92m:35240**]; English transl., Ukrainian Math. J. **43** (1991), no. 11, 1470–1476. [Zbl 779.35103]

[88] Sakhnovich, L. A., *Interpolation problems, inverse spectral problems and nonlinear equations,* in: Operator theory and complex analysis (Sapporo, 1991), 292–304, Oper. Theory Adv. Appl., 59, Birkhäuser, Basel, 1992. [MR **94h:47144**]

[89] Sakhnovich, L. A., *The method of operator identities and problems in analysis,* Algebra i Analiz **5** (1993), no. 1, 3–80. [MR **95f:47002**]; English transl., St. Petersburg Math. J., **5** (1994), no. 1, 1–69. [Zbl 823.47017]

[90] Sakhnovich, L. A., *Integral equations in the theory of stable processes,* Algebra i Analiz **4** (1992), no. 4, 225–238. [MR **94d:60118**]; English transl., St. Petersburg Math. J., **4** (1993), no. 4, 819–829. [Zbl 791.60103]

[91] Sakhnovich, L. A., *Inverse problems for equations systems,* in: Matrix and operator valued functions, 202–211, Oper. Theory Adv. Appl., 72, Birkhäuser, Basel, 1994. [MR **95k:34022**]

[92] Sakhnovich, L. A., *Nonlinear equations and inverse spectral problems on the axis,* in: Operator theory and boundary eigenvalue problems (Vienna, 1993), 244–255, Oper. Theory Adv. Appl., 80, Birkhäuser, Basel, 1995. [MR **96i:34184**]

[93] Sakhnovich, L. A., *Integral equations with difference kernels on finite intervals,* in: Oper. Theory Adv. Appl., 84. Birkhäuser Verlag, Basel, 1996. [MR **97e:45001**]

[94] Sakhnovich, L. A., *Interpolation Theory and its Applications,* Mathematics and its Applications, 428. Kluwer Academic Publishers, Dordrecht, 1997.

[95] Sakhnovich, L. A., *Weyl-Titchmarsh functions of the canonical periodical system of differential equations,* in: Topics in interpolation theory (Leipzig, 1994), 463–479, Oper. Theory Adv. Appl., 95, Birkhäuser, Basel, 1997. [MR **98k:34132**]

[96] Shilov, G. E., *Generalized Functions and Partial Differential Equations,* Authorized English edition revised by the author, Seckler Gordon and Breach, Science Publishers, Inc., New York-London-Paris 1968. [MR **37** #5694]

[97] Szegö, G., *Orthogonal Polynomials,* Fourth edition. American Mathematical Society, Colloquium Publications, Vol. XXIII. American Mathematical Society, Providence, R.I., 1975. [MR **51** #8724]

[98] Zakharov, V. E., and Shabat, A. B., *A scheme for integrating the nonlinear equations of mathematical physics by the method of the inverse scattering problem. I,* Funkcional. Anal. i Priložen., **8** (1974), no. 3, 43–53. [MR **58** #1768]; English transl., Funct. Anal. Appl., **8** (1974), no. 3, 226–235. [Zbl 303.35024]

[99] Zolotarev, V., *La Factorisation des Fonctions*, preprint, Université Pierre et Marie Curie, 1–35, 1983.

[100] Zygmund, A., *Trigonometric Series,* Vol. I, II. Reprint of the 1979 edition. Cambridge Mathematical Library, Cambridge University Press, Cambridge-New York, 1988. [MR **89c:42001**]

Index

MIX
Papier aus verantwortungsvollen Quellen
Paper from responsible sources
FSC® C105338

If you have any concerns about our products,
you can contact us on
ProductSafety@springernature.com

In case Publisher is established outside the EU,
the EU authorized representative is:
Springer Nature Customer Service Center GmbH
Europaplatz 3, 69115 Heidelberg, Germany

Printed by Libri Plureos GmbH
in Hamburg, Germany